AF691692

Introduction to FORENSIC CHEMISTRY

Introduction to FORENSIC CHEMISTRY

Jack Wilson
B.B. Long

Kruger Brentt
Publishers
2025

Kruger Brentt Publishers UK. LTD.
Company Number 9728962

Regd. Office: 68 St Margarets Road, Edgware, Middlesex HA8 9UU

© 2025 AUTHORS

ISBN: 978-1-78715-441-4

Disclaimer:

Every possible effort has been made to ensure that the information contained in this book is accurate at the time of going to press, and the publisher and author cannot accept responsibility for any errors or omissions, however caused. No responsibility for loss or damage occasioned to any person acting, or refraining from action, as a result of the material in this publication can be accepted by the editor, the publisher or the author. The Publisher is not associated with any product or vendor mentioned in the book. The contents of this work are intended to further general scientific research, understanding and discussion only. Readers should consult with a specialist where appropriate.

Every effort has been made to trace the owners of copyright material used in this book, if any. The author and the publisher will be grateful for any omission brought to their notice for acknowledgement in the future editions of the book.

All Rights reserved under International Copyright Conventions. No part of this publication may be reproduced, stored in a retrieval system, or transmitted in any form or by any means, electronic, mechanical, photocopying, recording or otherwise without the prior written consent of the publisher and the copyright owner.

For information on all our publications visit our website at http://krugerbrentt.com/

PREFACE

Forensic chemistry is a critical discipline that intersects the fields of chemistry and criminal justice, playing an essential role in the investigation and prosecution of crimes. This scientific domain involves the application of chemical principles and techniques to analyze physical evidence collected from crime scenes, enabling forensic scientists to provide vital information that can aid in solving cases.

The primary objective of forensic chemistry is to identify and quantify substances present in samples, including drugs, toxins, explosives, and trace materials, which can link suspects to criminal activities or exonerate innocent individuals.

In the laboratory, forensic chemists employ a variety of analytical methods, such as chromatography, mass spectrometry, and spectroscopy, to conduct detailed examinations of evidence. Chromatography is particularly useful for separating complex mixtures, allowing chemists to identify individual components within a sample.

Mass spectrometry helps in determining the molecular weight and structure of substances, providing insights into their identity. Spectroscopy techniques, including infrared and ultraviolet-visible spectroscopy, enable the characterization of materials based on their interaction with light. These analytical tools are essential in forensic investigations, as they provide quantitative and qualitative data that can be critical for legal proceedings.

One of the most significant applications of forensic chemistry is in toxicology, where chemists analyze biological samples, such as blood or urine, to detect the presence of drugs, alcohol, or poisons.

This information can be crucial in cases involving overdose, impaired driving, or poisoning. Additionally, forensic chemistry plays a vital role in arson investigations by analyzing fire debris to identify accelerants used to start fires. The examination of trace evidence, such as fibers, gunshot residue, or paint, also falls under the purview of forensic chemistry, providing valuable connections between suspects, victims, and crime scenes.

Moreover, forensic chemistry is continually evolving, incorporating advancements in technology and methodology to enhance the accuracy and reliability of analyses. The development of new techniques, such as laser ablation and nanotechnology, has expanded the capabilities of forensic chemists, allowing for more sensitive and specific analyses of minute quantities of evidence. In summary, forensic chemistry is an indispensable component of the criminal justice system, bridging scientific inquiry with legal processes to ensure that justice is served.

Its ability to analyze and interpret complex evidence makes it a cornerstone of modern forensic investigations, ultimately contributing to the resolution of criminal cases and the protection of society.

Jack Wilson

B.B. Long

CONTENTS

CHAPTER 1

AN INTRODUCTION TO FORENSIC CHEMISTRY AND PHYSICAL EVIDENCE

Forensic chemistry plays a pivotal role in the field of forensic science by applying chemical principles and techniques to the investigation of crime scenes, analyzing physical evidence, and contributing to the pursuit of justice. The discipline integrates diverse aspects of chemistry, including organic, inorganic, analytical, and physical chemistry, to help law enforcement agencies and legal systems decipher complex cases. Physical evidence, which includes anything tangible collected at crime scenes or from suspects and victims, is analyzed to establish connections, reconstruct events, and confirm or refute theories.

1.1 THE ROLE OF FORENSIC CHEMISTRY IN CRIMINAL INVESTIGATIONS

Forensic chemistry is indispensable in criminal investigations. Chemists analyze substances found at crime scenes—such as drugs, explosives, toxic chemicals, blood, and other biological materials—to identify their composition and origin. Forensic chemists work closely with forensic pathologists, toxicologists, and law enforcement to solve criminal cases by interpreting chemical data.

The work of forensic chemists extends beyond the mere identification of substances. They often provide critical insights into how evidence was created, altered, or transferred during the commission of a crime. Their analytical findings are often presented in court to support prosecution or defense arguments. Forensic chemists may also determine the time frames in which a substance was used or a crime occurred, based on chemical reactions or degradation rates.

1.2 PHYSICAL EVIDENCE: DEFINITION AND IMPORTANCE

Physical evidence is any object or material relevant to a criminal investigation that can establish links between crimes, suspects, and victims. It serves as a tangible record of events, behaviors, and interactions, and can range from fingerprints to chemical residues, weapon fragments, and biological samples like blood and saliva. Forensic chemists use advanced analytical techniques to examine physical evidence, ensuring that each item is scrutinized thoroughly for clues.

Physical evidence is critical because it can either corroborate witness testimony or provide objective information that contradicts it. In cases where witnesses are unavailable or unreliable, physical evidence becomes the primary source of truth. Its proper collection, preservation, and analysis can lead to accurate reconstructions of criminal activities, ultimately influencing court decisions.

1.3 TYPES OF PHYSICAL EVIDENCE

Forensic chemistry deals with a wide array of physical evidence, each requiring specific methods for analysis. The major types of physical evidence include:

1.3.1 Biological Evidence

Biological evidence comprises any material derived from living organisms, such as blood, semen, hair, saliva, and tissue samples. Forensic chemists often conduct DNA analysis to match biological samples from crime scenes with those of suspects or victims.

This type of evidence is particularly powerful due to the uniqueness of an individual's DNA profile. In addition to identifying individuals, biological evidence can also determine the presence of drugs, alcohol, or poisons in the system at the time of death or during the crime.

Forensic biology is the application of biological principles and techniques in the investigation of criminal and civil cases. Forensic biology is primarily concerned with analyzing biological and serological evidence in order to obtain a DNA profile, which aids law enforcement in the identification of potential suspects or unidentified remains. This field encompasses various sub-branches, including forensic anthropology, forensic entomology, forensic odontology, forensic pathology, and forensic toxicology.

History

The first recorded use of forensic procedures dates back to the 7th century when the concept of using fingerprints as a means of identification was first established. By the end of the 7th century, forensic procedures were being used to determine the guilt of criminals.

An early pioneer in criminal identification through biology was Alphonse Bertillon, also known as the "father of criminal identification". In 1879, he introduced a scientific approach to personal identification by developing the science of anthropometry. Anthropometry involves the use of a series of body measurements to distinguish human individuals.

Karl Landsteiner, in 1901, introduced the categorization of human blood into groups: A, B, AB, and O. From this discovery, blood typing, became a fundamental tool in forensic science.

After this, advancements were made which contributed to the ease of using and detecting blood found at crime scenes, expanding the use of blood analysis in forensic biology. Leone Lattes discoverd a method to determine the blood group of dried bloodstains in 1915. Subsequently, Albrecht H.O., a German chemist, developed luminol in 1928, which is used to detect trace blood stains at crime scenes.

Alec Jeffreys developed DNA fingerprinting in 1984, which examines variations in DNA that can identify individuals. This has become emminently useful not only in forensic science, but also in resolving paternity and immigration disputes.

In 1983, Kary B. Mullis expanded the use of DNA profiling by developing PCR (polymerase chain reaction), which amplifies DNA segments in-vitro, even in trace amounts. DNA samples found in crimes scenes are often found in minute amounts and degraded states, and sometimes mixed with various body fluids from multiple individuals. Using PCR, these DNA samples can be amplified for analysis when they otherwise would be useless. Beyond forensics, PCR has made an impact on a wide range of fields, including disease diagnosis and virus detection.

DNA Analysis

DNA, or deoxyribonucleic acid, is one of the most popular pieces of evidence to recover at a crime scene. Evidence containing DNA is regarded as *biological evidence*, and is recognized as the "golden standard" in forensic science.

DNA analysis has numerous applications, such as paternity testing, identification of unknown human remains, breakthroughs in cold cases, as well as connecting suspects and/or victims to a piece (or pieces) of evidence, a scene, or another person (either a victim or suspect). Nuclear DNA evidence can be recovered from blood, semen, saliva, epithelial cells and hair (provided the root is still intact). Furthermore, Mitochondrial DNA (mtDNA) can be recovered from the shaft of hair, bone, and the roots of teeth.

To be used, biological evidence must be initially visually recognized at the crime scene. To aid in this, alternative light sources, or an Advanced Light Source (ALS), are used. Once a potential source is identified, presumptive tests are conducted to establish if there is a specified biological presence (semen, saliva, blood, urine, etc.). If positive, samples are collected and submitted for analysis to a laboratory, where confirmatory tests and further tests are performed.

For most forensic DNA samples, STR analysis of autosomal short tandem repeats (STR) is performed in an attempt to individualize the sample to one person with a high degree of statistical confidence. Here, STR markers for autosomal STR are used in forensic DNA typing to track down the missing, verify family connections, and potentially connect suspects to crime sites.

Laboratory analysis of DNA evidence involves the sample DNA being extracted, quantified, amplified, and visualized. There are several methods of DNA extraction possible including organic (phenol-chloroform) extraction, Chelex extraction, and differential extraction.

Quantitation is commonly conducted using a form of the polymerase chain reaction, known as real-time PCR, quantitative PCR (qPCR). qPCR is the preferred method of DNA quantitation for forensic cases because it is very precise, human-specific, qualitative, and quantitative. This technique analyses changes in fluorescence signals of amplified DNA fragments between each PCR cycle without needing to pause the reaction or open the temperature-sensitive PCR tubes. In addition to the components necessary for a standard PCR reaction (i.e. template DNA, carefully designed forward and reverse primers, DNA polymerase [usually *Taq*], dNTPs, and a buffer solution containing Mg2+), qPCR reactions involve fluorescent dye-labelled probes that complement and anneal to the DNA sequence of interest that lies between the two primers.

A "reporter" (R) dye is attached at the 5' end of the fluorescent probe, while a "quencher" (Q) dye is attached at the 3' end. Before the DNA strands are extended by the polymerase, the reporter and quencher are close enough in space that no fluorescence is detected by the instrument (the quencher completely absorbs/masks the fluorescence of the reporter). As the polymerase begins to extend the strand, the 5' end of the probe is degraded by the polymerase due to its exonuclease activity.

The reporter dye is released from the 5' end. It is no longer quenched, thus enabling fluorescence detection. A graph is constructed for the sample DNA comparing the presence of fluorescence (y-axis) to cycle number (x-axis) of the qPCR process. This is then compared to a standard curve of the cycle fluorescence threshold (y-axis) versus the log of known DNA concentrations (x-axis). By comparing the sample data to the standard curve, one may extrapolate the DNA concentration in the sample, which is essential to move forward with PCR amplification and capillary electrophoresis to obtain a DNA profile.

DNA profiles are produced as an electropherogram. The obtained profile can be compared to known samples in CODIS to identify a possible suspect. Based on known frequencies of the genotype found in the DNA profile, the DNA analyst may place a statistical measure of confidence on DNA match.

Mitochondrial DNA Analysis

Mitochondrial DNA (mtDNA) is used instead of nuclear DNA when forensic samples have been degraded, are damaged, or are in very small quantities. In many cases, these may be older human remains, sometimes ancient, and the only options for DNA collection are the body's bone, teeth, or hair.

mtDNA can be extracted from degraded samples since its presence in cells is much higher than nuclear DNA. There can be more than 1,000 copies of mtDNA in a cell, while there are only two copies of nuclear DNA. Nuclear DNA is inherited from both the mother and the father but mtDNA is passed down from only the mother to all of her offspring. Due to this type of inheritance, mtDNA is useful for identification purposes in forensic work but can also be used for mass disasters, missing persons cases, complex kinship, and genetic genealogy.

The main advantage of using mtDNA is its high copy number. However, there are a few disadvantages of using mtDNA as opposed to nuclear DNA. Since mtDNA is inherited maternally and passed to each offspring, all members of the maternal familial line will share a haplotype. A haplotype "is a group of alleles in an organism that are inherited together from a single parent". Sharing this haplotype among family members can cause an issue in forensic samples because these samples are often mixtures that contain more than one DNA contributor. The convolution and interpretation of mtDNA mixtures is more difficult than that of nuclear DNA, and some laboratories choose not to attempt the process Since mtDNA does not recombine, the genetic markers are not as diverse as autosomal STRs are in the case of nuclear DNA. Another issue is that of heteroplasmy — when an individual has more than one type of mtDNA in their cells. This can cause an issue in interpreting data from questioned forensic samples and known samples that contain mtDNA. Having adequate knowledge and understanding of heteroplasmy can help ensure successful interpretation.

There are some ways to improve success of mtDNA analysis. Preventing contamination at all testing stages and using positive and negative controls is a priority. In addition, the use of mini-amplicons can be beneficial. When a sample of mtDNA is severely degraded or has been obtained from an ancient source, the use of small amplicons can be used to improve the success of amplification during PCR. In these cases primers amplifying smaller regions of HV1 and HV2 in the control region of mtDNA are used. This process has been referred to as the 'ancient DNA' approach.

The first use of mtDNA as evidence in court was in 1996 in *State of Tennessee v. Paul Ware*. There was only circumstantial evidence otherwise against Ware so the admittance of mtDNA from hairs found in the victim's throat and at the scene were key to the case.

In 2004, with the help of the National Center for Missing and Exploited Children and ChoicePoint, mtDNA was used to solve a 22-year-old cold case where the nuclear DNA evidence was not originally strong enough. After mtDNA analysis, Arbie Dean Williams was convicted of the murder of 15-year-old Linda Strait, which had occurred in 1982.

In 2012, mtDNA evidence allowed investigators to establish a link in a 36-year-old investigation into the murders of four Michigan children. Hair fibers found on the bodies of two of the children were tested and the mtDNA found to be the same for each sample. For the investigators this was a big break because it meant that the murders were likely connected.

Disciplines

Forensic Anthropology

Anthropology is applied to forensics most regularly through the collection and analysis of human skeletal remains.

The primary goals of anthropological involvement include identification and aiding in scene reconstruction by determining details regarding the circumstances of the victim's death. In cases where conventional techniques are unable to determine the identity of the remains due to the lack of soft tissue, anthropologists are required to deduce certain characteristics based on the skeletal remains. Race, sex, age and possible ailments can often be determined through bone measurements and looking for clues throughout the skeletal structure. This becomes necessary when conventional methods that use soft tissue fail to establish the identity of remains.

Forensic Botany

Forensic botany is the application of plant science to legal investigations. It involves the study of plant material, such as leaves, seeds, pollen, and other plant properties, to gather evidence that can aid in criminal or civil proceedings. The identification of plant material is crucial in forensic botany, as it can provide a link between an individual and a crime scene, pinpoint the geographical location of missing bodies, or establish the post-mortem interval (PMI) of a human skeleton.

Forensic botany can also help investigators determine the cause of death in cases where plant toxins are involved. For example, the presence of certain plant species in the stomach contents of a deceased individual may indicate accidental or intentional poisoning.

In addition to identifying plant material, forensic botanists may also analyze soil samples for traces of plant material, which can provide valuable information about the environment in which a crime occurred. The analysis of plant material and soil samples can be performed through various techniques, including optical microscopy, scanning electron microscopy, and DNA analysis.

Subspecialties in Forensic Botany

Subdisciplines within forensic botany include:

- **Forensic Palynology (Study of Pollen and Spores):** Palynology can produce evidence of decomposition time, location of death or the time of year.

- **Bryology (Study of Bryophytes):** Bryology is the easiest to find evidence since bryophyte (a species of plants) attaches to shoes and clothes easily. Bryophytes are useful since even if they are ripped apart or broken down, DNA can still be analyzed.
- Dendrochronology (study of the growth of rings of trees stems and roots).
- Lichenology (study of lichens).
- Mycology (study of fungi).

Forensic Ornithology

Forensic ornithology is the application of scientific techniques to the examination and identification of bird remains for legal purposes. This field of study can aid in investigations related to wildlife crimes, such as poaching, smuggling, and illegal trade of birds and their feathers.

Feathers are one of the most important pieces of evidence used in forensic ornithology. Each bird species has unique feather characteristics that can be observed at both macroscopic and microscopic levels. These characteristics include the size, shape, color, and pattern of the feather, as well as the arrangement and structure of the barbs and barbules. By examining these features, a forensic ornithologist can determine the species of bird to which a feather belongs.

Other types of bird remains can also be identified through forensic ornithology. Bones, for example, can be analyzed to determine the species of bird, as well as the age and sex of the individual. Blood samples can also be used to identify bird species through DNA analysis.

Forensic ornithology can be used in a variety of contexts, including criminal investigations, wildlife management, and conservation efforts. By providing accurate identification of bird remains, this field of study can help to bring perpetrators of wildlife crimes to justice and protect endangered bird species.

Forensic Odontology

Forensic odontology, also known as forensic dentistry, is the application of dental science to legal matters. It is a specialized field that has been instrumental in helping law enforcement detect and solve cases in criminal and civil proceedings.

The use of forensic odontology became more popular in the 1960s with the establishment of the first instructional program in the United States at the Armed Forces Institute of Pathology. Since then, forensic odontology has become widely known by both dental and law enforcement professionals.

Forensic odontologists are dental professionals who use their expertise to establish a person's identity, interpret injuries in the oral and perioral regions, analyze and compare bite marks, and assist forensic pathologists in determining the cause of death if there is a possible contributing dental condition.

Dental evidence is a valuable tool in establishing human identity by comparing the dental features of a deceased person with antemortem dental records. Forensic odontologists can also assist in age estimation in living and deceased persons, which can be useful in cases where the identity of a person is unknown.

Forensic Pathology

Forensic pathology is a specialized field within forensic science that focuses on the examination of individuals who have died suddenly, unexpectedly, or violently to determine the cause and manner of death. A forensic autopsy is a postmortem examination of the body and analysis of bodily fluids to provide information on the cause of death, manner of death, and mechanism of injury.

A forensic pathologist is a medical doctor who has extensive knowledge and expertise in both trauma and disease. They are responsible for performing autopsies and applying their knowledge of the human body and possible internal and external injuries to determine the cause and manner of death. The information obtained from an autopsy can greatly assist investigative efforts and scene reconstruction. A forensic pathologist may also collect evidence from the body, such as trace evidence or biological fluids, that can be used in criminal investigations, and testify in court as expert witnesses regarding their findings.

Forensic Toxicology

Forensic toxicology is an interdisciplinary field that applies principles and methods from toxicology, analytical chemistry, pharmacology, and clinical chemistry to aid in medical or legal investigations of death, poisoning, and drug use. The primary objective of forensic toxicology is the accurate detection, identification, and interpretation of chemicals and their metabolites in biological samples for the purpose of providing objective evidence to support medical or legal decisions. The results of forensic toxicology analyses can be used to determine the cause and manner of death, assess the role of drugs or chemicals in impairment or toxicity, and provide evidence in criminal or civil proceedings. The field of forensic toxicology requires a thorough understanding of the pharmacokinetics and pharmacodynamics of drugs and chemicals, as well as the analytical techniques used to detect and quantify them in biological matrices.

Forensic Microbiology

Forensic microbiology has become an increasingly promising area of research with the recent advances in massive parallel sequencing, also known as next-generation sequencing. This technology has enabled the analysis of microorganisms for various applications in forensic science, including biocrime, bioterrorism and epidemiology.

Microorganisms can serve as valuable sources of evidence in criminal cases, including:

- **Clarifying Causes of Death:**
  - Drowning;
  - Toxicology;
  - Hospital-acquired infections;
  - Sudden infant death; and
  - Shaken baby syndrome.
- **Assisting in Human Identification through the Analysis Of:**
  - Skin flora;
  - Hair microbiome; and
  - Body fluid microbiome.
- **Geolocation Purposes by Analyzing:**
  - Soil microbiome; and
  - Microbiomes of bodies of water.
- **Estimation of the Post-Mortem Interval Via:**
  - The thanatomicrobiome, the community of microorganisms that colonize a body after death; and
  - Epinecrotic microbial communities, the microorganisms that colonize the surface of a decomposing body.

By analyzing the stage of decay from bacterial decomposition or the bacterial succession patterns, scientists can estimate the time elapsed since death.

Bioterrorism and Epidemiology

Bioterrorism refers to the deliberate use of biological agents as weapons of warfare. These agents, which can be natural or genetically modified microorganisms, are intentionally disseminated to cause illness, death, or harm to humans, animals, or plants. Regardless of their origin, these biological weapons, which can be viruses, bacteria, or fungi, are highly infectious and pose a significant threat.

Forensic microbiology plays a crucial role in the study of epidemiology. By examining microorganisms obtained from infected individuals, scientists can determine the source of infection, identify the type of infection present, and analyze the mutation pattern of the microorganism. Forensic microbiologists compare the microorganisms isolated from infected individuals to known sources of infectious pathogens to identify the cause of an outbreak.

It is important to note that biological agents used as weapons are often found in the environment, making it challenging to determine whether an infection is accidental or the result of a deliberate attack. One of the most notable cases of bioterrorism in recent history involved the mailing of at least four anthrax-containing envelopes in the United States in September and October 2001. This incident resulted in 11 people contracting inhalation anthrax, leading to five fatalities, while another 11 individuals were afflicted with cutaneous anthrax. Additionally, 31 persons tested positive for exposure to *Bacillus anthracis* spores.

However, advancements in PCR and whole-genome sequencing enabled scientists to collaborate with the FBI to identify the source of the letter spores. The combination of forensic microbiology and modern technology is essential in identifying and preventing bioterrorism attacks.

Differentiating a Biowarfare Attack from a Normal Epidemiology Outbreak

When investigating a possible bioterrorist or biowarfare attack, the epidemiologic approach is unique from a typical epidemiologic investigation. The process begins by confirming that there has been an outbreak using laboratory and clinical evidence. Once the number of cases and definition of the attack has been established, the outbreak can be characterized by analyzing the time, place, and person affected. This information is critical in identifying the source of the outbreak. By gathering data on cases over time, an epidemic curve can be created. The pattern of the disease is important in distinguishing between a natural outbreak and an intentional attack. In the case of a bioterrorism attack, the source is most likely a single point, with everyone coming into contact with the agent at the same time. Other factors that are investigated to determine if the outbreak is the result of a biological attack include a large epidemic, more severe disease than expected for a given pathogen, an uncommon disease for a specific area, and multiple simultaneous epidemics of different diseases.

Post-Mortem Microbial Analysis

Post-mortem microbiology is a field that aims to detect unexpected infections causing sudden deaths, confirm clinically suspected but unproven infections, evaluate the efficacy of antimicrobial therapy, identify emergent pathogens, and recognize medical errors. Additionally, the analysis of the thanatomicrobiome may help estimate the post-mortem interval. Currently, extensive research is being conducted to determine if there is a consistent microbial decomposition "clock" that could be used by itself or in conjunction with other techniques, such as forensic entomology, to help estimate postmortem intervals.

One research group has made significant progress in describing such a microbial clock and believes they are within two to five years of testing it in a real crime scene scenario. However, if a reliable and consistent microbial clock is determined to exist, it remains to be seen whether it will pass scientific and legal muster. A judge would also have to determine that the microbial clock meets the standard for admission of expert testimony.

Forensic Limnology

Forensic limnology is the application of limnology, the study of inland waters, to forensic science. In cases involving a body of water at or near the scene of a crime, a sample of the water can be extracted and analyzed to identify the presence and composition of microorganisms, which can act as a form of trace evidence. One such microorganism are diatoms, a type of microalgae that vary in shape and are unique to specific bodies of water. By analyzing the diatom composition in a sample of water, investigators can establish if a person or questioned piece of evidence has been in contact with a specific body of water. This is because diatoms are specific to certain bodies of water, and if a sample contains diatoms found only in a specific body of water, it can be used as evidence to link a person or object to that location. Forensic limnology can be used in conjunction with other forensic techniques to provide a more comprehensive analysis of a crime scene.

Current Issues

Sexual Assault Kit Backlog

As DNA is a critical form of evidence in the investigation of cases of sexual violence, the backlog of untested sexual assault kits (SAKs), also known as a rape kit, greatly affects the successful identification and prosecution of the perpetrators of these crimes.

According to RAINN (Rape, Abuse & Incest National Network), the largest anti-sexual violence organization in the United States, the backlog is the result of both a failure of law enforcement to send collected kits to forensic laboratories for analysis, and a lack of resources within these labs to process the kits effectively. In the absence of adequate funding, many districts would rather dedicate their funds to homicides or more high-profile cases, and cases of sexual violence are often swept to the side. With this, as the SAKs remain in storage, the prevalence of the issue increases, especially as more and more kits are being found each year.

Cold Cases

With the considerable advances in DNA analysis, old, open cases with intact evidence can be examined for biological evidence. New profiles are uploaded to CODIS every day so the base population to search and compare to increases. Biological testing for cold cases, specifically homicides, encounters similar roadblocks as the SAKs - lack of funds or the DNA samples have not been properly stored; thus too much degradation has occurred for viable analyses.

1.3.2 Drug and Chemical Evidence

Drug and chemical evidence are critical components in forensic chemistry, playing an essential role in criminal investigations and legal proceedings. Forensic chemists analyze these substances to identify their composition, determine their origin, and assess their potential impact in criminal cases.

This type of evidence is frequently encountered in investigations related to drug trafficking, poisoning, substance abuse, arson, and violent crimes, making its accurate analysis crucial for establishing connections between suspects, victims, and criminal activities.

Drug and chemical evidence form a cornerstone of forensic chemistry. The precise identification and quantification of these substances can provide critical information that helps solve crimes, identify perpetrators, and exonerate the innocent. The ongoing evolution of drugs and chemical compounds continues to push forensic science to adopt newer, more sophisticated technologies and methodologies to ensure accurate and timely analyses.

Types of Drug and Chemical Evidence

The primary forms of drug and chemical evidence include:

- **Illicit Drugs:** Substances such as narcotics (heroin, cocaine), hallucinogens (LSD, psilocybin), stimulants (methamphetamine, ecstasy), and opioids are commonly encountered in forensic laboratories. These drugs may be found in raw form, processed tablets, powders, or mixed in other compounds.
- **Controlled Prescription Drugs:** Many cases involve the illegal distribution or misuse of prescription medications like opioids (oxycodone, fentanyl), sedatives (benzodiazepines), or stimulants (amphetamine-based drugs).
- **Poisons and Toxic Substances:** Chemicals like cyanide, arsenic, and various industrial or household chemicals are often tested in suspected poisoning cases. The analysis helps determine if a toxic substance was involved in a crime, especially in cases of homicide or suicide.
- **Explosives and Flammable Chemicals:** In arson and bombing investigations, forensic chemists analyze residues from the scene to detect the presence of accelerants (such as gasoline) or explosive chemicals (like ammonium nitrate or TNT).
- **Trace Evidence:** Even small quantities of drugs or chemicals found as residues on clothing, surfaces, or containers can be used to establish connections between a suspect and a crime scene.

Analytical Techniques for Drug and Chemical Evidence

Forensic chemists utilize several advanced analytical techniques to identify and quantify drugs and chemicals in evidence:

- **Gas Chromatography-Mass Spectrometry (GC-MS):** This is one of the most common techniques used to separate and identify components in a sample. GC separates the chemicals based on their volatility, while MS provides detailed molecular information, allowing for precise identification of substances.
- **Liquid Chromatography-Mass Spectrometry (LC-MS):** This technique is especially useful for analyzing non-volatile substances, including certain drug formulations or complex biological samples.
- **Fourier Transform Infrared Spectroscopy (FTIR):** FTIR measures how a sample absorbs infrared light, providing a chemical fingerprint that can help identify organic substances, including drugs.
- **Ultra-Violet (UV) Spectroscopy:** UV spectroscopy is used for the detection and quantification of specific drugs by analyzing how they absorb ultraviolet light. This method is commonly used for screening substances in forensic laboratories.
- **Color Tests and Chemical Spot Tests:** While not as conclusive as chromatography or spectroscopy, these quick tests provide presumptive results by causing color changes when a drug or chemical reacts with certain reagents. They are often used at crime scenes for initial screening.

Legal Implications and Challenges

Forensic analysis of drug and chemical evidence plays a crucial role in legal contexts. The results must meet strict standards to be admissible in court, including the proper chain of custody, accurate documentation, and method validation. The integrity of the evidence and the objectivity of the forensic chemist are paramount to ensuring justice is served.

Challenges in drug and chemical evidence analysis include the emergence of new synthetic drugs or altered substances designed to bypass legal controls. Forensic laboratories must stay updated with new testing methods and reference materials to accurately identify these novel compounds.

1.3.3 Explosive and Fire Debris Evidence

Explosive and fire debris evidence play a pivotal role in forensic chemistry, especially in cases involving arson, bombings, and accidental explosions. The analysis of such evidence is crucial in determining the cause and origin of a fire or explosion, identifying accelerants or explosive materials, and linking suspects to the crime scene.

Forensic chemists are responsible for carefully collecting, preserving, and analyzing debris to extract valuable information that can aid in solving these cases. Explosive and fire debris evidence analysis is a vital aspect of forensic chemistry. Through the application of advanced analytical techniques, forensic chemists can identify explosive materials and accelerants, reconstruct the events leading to an explosion or fire, and provide critical evidence for investigations. The meticulous analysis of such evidence not only assists in solving crimes but also helps ensure that justice is served in legal proceedings.

Explosive Evidence

Explosive evidence refers to residues or remnants from explosive materials used in bombings or other intentional detonations. Forensic chemists analyze this evidence to identify the type of explosive used and to reconstruct the sequence of events leading to the explosion.

- **Types of Explosives:**
  - *Low Explosives:* These materials, such as black powder or smokeless powder, burn rapidly rather than detonating. They are often used in improvised explosive devices (IEDs) and pyrotechnics.
  - *High Explosives:* Substances like TNT, RDX, and PETN are capable of detonating at extremely high velocities. They are commonly used in military applications, large-scale bombings, and commercial demolitions.
  - *Collection of Evidence:* The collection of explosive evidence requires meticulous care to preserve residues that can provide clues about the materials used. Fragments of the explosive device, traces of powder, or chemical residues can be collected from the scene. Improper handling can contaminate evidence, so strict chain-of-custody procedures must be followed.
- **Analysis Techniques:**
  - *Gas Chromatography-Mass Spectrometry (GC-MS):* This method is widely used to identify explosive residues in the debris. GC-MS can separate and analyze volatile components, helping identify specific explosive materials.
  - *Infrared Spectroscopy (IR):* IR spectroscopy helps detect the functional groups present in the explosive material, aiding in its identification.
  - *Ion Chromatography (IC):* IC is employed to detect and quantify ionic species, such as nitrates or perchlorates, which are common components of explosives.
  - *Scanning Electron Microscopy (SEM):* SEM coupled with energy dispersive X-ray spectroscopy (EDX) is used to examine tiny particles of explosives, providing both physical and chemical information.

Fire Debris Evidence

Fire debris evidence refers to materials recovered from a fire scene that may contain traces of accelerants or other substances used to initiate or fuel a fire. Forensic chemists analyze these samples to determine whether a fire was accidental or intentionally set (arson).

- **Accelerants:** Accelerants are flammable or combustible liquids (e.g., gasoline, kerosene, lighter fluid) that are commonly used to start or spread fires. Their presence in fire debris can indicate arson. However, some household materials may also contain flammable substances, so forensic chemists must differentiate between legitimate sources and criminal use.
- **Collection of Fire Debris:** Fire debris must be collected from areas where accelerants are suspected to have been used. Samples are typically placed in airtight containers to prevent the loss of volatile compounds. This ensures that any residual accelerants can be identified during analysis.
- **Analysis Techniques:**
  - *Headspace Gas Chromatography (HS-GC):* This technique is used to detect volatile compounds in fire debris. The debris sample is heated, causing volatile compounds to vaporize. These vapors are then analyzed by GC to identify accelerants.
  - *Gas Chromatography-Mass Spectrometry (GC-MS):* GC-MS is a powerful tool for identifying complex mixtures of accelerants, even in small quantities. It can separate and identify the individual components present in the debris.
  - *Portable Gas Detectors:* In some cases, portable gas detectors are used on-site to screen for the presence of accelerants. These instruments provide rapid results, allowing investigators to prioritize certain areas of the scene for detailed analysis.
- **Challenges in Fire Debris Analysis:** Fire debris analysis can be complicated due to the destructive nature of fire. High temperatures may degrade chemical evidence, making it more difficult to identify accelerants. Additionally, certain materials (e.g., plastics, furniture polish) may produce volatile compounds that can mimic the presence of accelerants, leading to potential false positives.

Legal and Investigative Significance

The analysis of explosive and fire debris evidence is crucial in both criminal and civil cases. It helps investigators determine whether a fire or explosion was accidental or deliberate and can link a suspect to the crime. The evidence must be handled with the utmost care to preserve its integrity for court proceedings, as forensic chemists often serve as expert witnesses to testify about their findings.

In bombing cases, identifying the specific explosive used can provide insights into the perpetrator's resources, capabilities, and potential connections to other criminal activities. In arson investigations, detecting accelerants in fire debris can indicate intentional fire-setting and help pinpoint the origin of the fire.

1.3.4 Toxicological Evidence

The discipline of evidence-based toxicology (EBT) strives to transparently, consistently, and objectively assess available scientific evidence in order to answer questions in toxicology, the study of the adverse effects of chemical, physical, or biological agents on living organisms and the environment, including the prevention and amelioration of such effects. EBT has the potential to address concerns in the toxicological community about the limitations of current approaches to assessing the state of the science. These include concerns related to transparency in decision making, synthesis of different types of evidence, and the assessment of bias and credibility. Evidence-based toxicology has its roots in the larger movement towards evidence-based practices.

By analogy to evidence-based medicine (EBM), the umbrella term evidence-based toxicology (EBT) has been coined to group all approaches intended to better implement the above-mentioned evidence-based principles in toxicology in general and in toxicological decision-making in particular. Besides systematic reviews, the core evidence-based tool, such approaches include *inter alia* the establishment and universal use of a common ontology, justified design and rigorous conduct of studies, consistently structured and detailed reporting of experimental evidence, probabilistic uncertainty and risk assessment, and the development of synthesis methodology to integrate evidence from diverse evidence streams, e.g. from human observational studies, animal studies, in vitro studies and in silico modeling. A main initial impetus for translating evidence-based approaches to toxicology was the need to improve the performance assessment of toxicological test methods. The U.S. National Research Council (NRC) concurs that new means of assessment are needed to keep pace with recent advances in the development of toxicological test methods, capitalizing on enhanced scientific understanding through modern biochemistry and molecular biology.

A key tool in evidence-based medicine that holds promise for EBT is the systematic review. Historically, authors of reviews assessing the results of toxicological studies on a particular topic have searched, selected, and weighed the scientific evidence in a non-systematic and non-transparent way. Due to their narrative nature, these reviews tend to be subjective, potentially biased, and not readily reproducible. Two examples highlighting these deficiencies are the risk assessments of trichloroethylene and bisphenol A (BPA). Twenty-seven different risk assessments of the evidence that trichloroethylene causes cancer have come to substantially different conclusions. Assessments of BPA range from low risk of harm to the public to potential risks (for some populations), leading to different political decisions. Systematic reviews can help reducing such divergent views. In contrast with narrative reviews, they reflect a highly structured approach to reviewing and synthesizing the scientific literature while limiting bias. The steps to carrying out a systematic review include framing the question to be addressed; identifying and retrieving relevant studies; determining if any retrieved studies should be excluded from the analysis; and appraising the included studies in terms of their methodological quality and risk of bias. Ultimately the data should be synthesized across studies, if possible by a meta-analysis. A protocol of how the review will be conducted is prepared ahead of time and ideally should be registered and/or published.

Scientists have made progress in their efforts to apply the systematic review framework to evaluating the evidence for associations between environmental toxicants and human health risks. To date, researchers have shown that important elements of the framework established in evidence-based medicine can be adapted to toxicology with little change, and some studies have been attempted. Researchers using the systematic review methodology to address toxicological concerns include a group of scientists from government, industry, and academia in North America and the European Union (EU) who have joined together to promote evidence-based approaches to toxicology through the nonprofit Evidence-based Toxicology Collaboration (EBTC). The EBTC brings together the international toxicology community to develop EBT methodology and facilitate the use of EBT to inform regulatory, environmental and public health.

Background

Evidence-based approaches were first conceived as a means of anchoring policy decisions, not to current practices or the beliefs of experts, but to experimental evidence. Evidence-based medicine (EBM) was launched slightly later. Its rise as a distinct discipline is generally credited to the work and advocacy of Scottish epidemiologist Archie Cochrane. The Cochrane Collaboration named in his honor was launched at Oxford University in 1993 to promote evidence-based reviews of clinical medical literature. More recently, EBM expanded to encompass evidence-based health care (EBHC).

EBM/HC involves the conscientious, explicit, and judicious use of current best evidence in making decisions about the care of individual patients taking patients' preferences into account. Prior to EBM, medical decisions about diagnosis, prevention, treatment or harm were often made without a rigorous evaluation of the alternatives. Research in the 1970s and 1980s showed that different physicians regularly recommended different treatments and tests for patients with ailments that were essentially the same, and that large proportions of procedures being performed by physicians were considered inappropriate by the standards of medical experts.

EBM/HC supporters stress that while evidence always has been important to the practice of medicine, EBM/HC provides an enhanced approach of identifying, assessing, and summarizing evidence. EBT's supporters make a similar argument.

The idea of translating evidence-based approaches from medicine to toxicology has been percolating for two decades, with proponents in both medicine and toxicology. Three research papers published in 2005 and 2006 catalyzed what eventually became known as EBT by suggesting that EBM's established tools and concepts might serve as a prototype of evidence-based decision-making in toxicology.

Process and Progress

The First International Forum Toward Evidence-Based Toxicology was held in 2007. The forum was organized by the European Commission and attended by 170 scientists from more than 25 European, American, and Asian countries. The goal was to explore the available concepts of EBT, and to launch an initiative to formally implement evidence-based assessment methods in toxicology.

The starting point for the discussions were two research papers suggesting that the tools and concepts established in evidence-based medicine could serve as a prototype of evidence-based decision-making for evaluating toxicological data. Apparent fundamental differences between medicine and toxicology were carefully considered during these discussions. Forum participants attempted to bridge the two disciplines in order to make use of the accrued wisdom and apply this approach to toxicology.

The proceedings of this forum were published as a special issue in *Human & Experimental Toxicology*.

EBT's proponents include experts in EBM, public health, and toxicology who believe that EBT can help toxicologists to better serve the goals of health protection and safety assurance.

They argue that EBT's methodologies for collecting, appraising, and pooling evidence can help ensure that all available information on a given topic is evaluated in a transparent, unbiased, and reproducible manner. They contend that EBT's concept of the systematic review could prove particularly helpful for the standardization and quality assurance of novel methodologies for evaluating toxicity, as well as for their formal validation. In this regard, EBT may prove particularly useful for assessing the performance of newer non-animal "21st century" toxicology tools.

EBT can also help scientists integrate new toxicological test methods into test strategies being implemented across the globe.

In 2010, a group of EBT supporters joined together to convene a workshop titled "21st Century Validation for 21st Century Tools". The session on the potential for evidence-based approaches to assess the performance of the new generation of non-animal test methods inspired the formation of the EBTC. The EBTC was officially launched in the U.S. in 2011 at a Society of Toxicology conference and convened its first workshop in 2012. The EBTC's EU branch was officially opened during the 2012 Eurotox conference.

In 2014, the EBTC hosted a workshop on "The Emergence of Systematic Review and Related Evidence-based Approaches in Toxicology" with speakers representing US and European organizations that are implementing and promoting the use of systematic reviews for toxicological questions.

The experts noted that the structured approach of systematic reviews increases objectivity and transparency but also made clear that the approach requires a substantial time investment, which is a challenge to its more widespread adoption. Consequently, the participants called for close collaboration of interested organizations, which they determined to be a pre-requisite for the broad and efficient introduction of systematic reviews in toxicology.

Applications of EBT

Regulatory Decision-Making

Some scientists and policymakers would like EBT to help them combine information from various sources. Toxicological evidence can be assigned to evidence streams, sets of studies representing the same type or level of evidence, such as human (observational) studies, animal studies, in vitro or mechanistic studies.

EBT can be applied both within one evidence stream, and it is especially well-suited to be applied across multiple evidence streams. Regulators often designate one study as "the lead study", then use later studies as additional information. Many perceive this as unsatisfying, but objective approaches to combine study results are lacking. The EBM concept of the systematic review has promise for this application, and some structured reviews serve as forerunners for this approach.

Evaluating Effects of Environmental Exposures

The U.S. National Toxicology Program's Office of Health Assessment and Translation (OHAT) has started to use systematic review methodology for the program's evaluations. The first systematic review was completed in 2016, reviewing the effects of fluoride on learning and memory in animal studies. OHAT's approach is tailored to its mandate, but its seems especially appropriate for substances with substantial yet conflicting literature, and hence the need for systematic reviews to sort out somewhat confusing situations.

Causation

One application of EBT focuses on causation. It addresses the challenge of tracing a health effect back to a toxicant, such as lung cancer to smoking. This approach is similar to legal arguments Some experts warn that this approach could increase the evidence burden for proving causation, and thereby increase the difficulty involved in banning toxic substances.

Clinical Toxicology

Practitioners of clinical toxicology, which is concerned with the treatment of patients known to be exposed to toxic substances, are also beginning to use an EBM-style approach. Guidance documents based on this approach have already been published

21st Century Toxicology

The National Research Council's (NRC) landmark 2007 publication, "Toxicity Testing in the 21st Century", has also been an impetus for EBT. EBT provides new tools for assessing test method performance. Also, as the focus of 21st-century toxicology shifts from animal biology to human biology, EBT provides a method for comparatively evaluating the results gleaned from new methods of investigating the effects of chemical exposure. The Evidence-based Toxicology Collaboration has pioneered a number of projects aimed at applying EBT approaches and systematic reviews to test methods comparison.

Limitations and Challenges

The specific differences between toxicology and medicine/health care cause challenges for implementing EBT. Evidence-based methodology of clinical research has been focused on a single type of study—randomized, controlled clinical trials, which are a direct measure of the effectiveness of the health care intervention under scrutiny. In contrast, toxicology employs a variety of different kinds of studies in three distinct evidence streams: human (observational) studies, animal studies, and non-animal studies. Because human evidence is frequently lacking most evidence is obtained by using animal and non-animal models, which—by definition—is more difficult to generalize and extrapolate to humans.

This methodological heterogeneity complicates evidence integration within an evidence stream, such as when inconsistent evidence is obtained from different animal species, but even more so across evidence streams. Adding to the difficulty is the reality that much toxicological evidence, more so than in medicine and health care, is not readily accessible in the literature. Moreover, the role of expert judgment, especially in systematic reviews, needs to be clearly defined, as it is a common misperception that evidence-based approaches leave no room for it. Systematic reviews should strive to make expert judgments clear along with the scientific basis for those judgments in developing conclusions for a systematic review. Further issues to be worked out include exposures to multiple substances, the multitude of outcomes observed in some animal studies, and challenges in improving the experimental designs and reporting of studies.

1.3.5 Trace Evidence

Trace evidence occurs when objects make contact, and material is transferred. This type of evidence is usually not visible to the naked eye and requires specific tools and techniques to be located and obtained. Due to this, trace evidence is often overlooked, and investigators must be trained to detect it. When it comes to an investigation trace evidence can come in many different forms and is found in a wide variety of cases. This evidence can link a victim to suspects and a victim or suspect to the crime scene.

There are three general categories in which forensic science uses trace evidence. It can be used for investigative aids, associative evidence, and in-scene reconstructions. In terms of investigative aids, trace evidence can provide information to determine the origin of a sample and determine the manufacture date of the material, all of which can limit potential suspects in a case. Associative evidence can associate with or link victims or suspects to a crime scene.

For reconstructions, trace evidence can provide information to understand how a crime occurred or the events that occurred before the crime.

Importance

The importance of trace evidence in criminal investigations was shown by Edmond Locard in the early 20th century, with his exchange principle, that every contact leaves a trace.

This statement can then be expanded by stating trace evidence must first be located and recorded before it can be recovered and analyzed. Since then, forensic scientists use trace evidence to reconstruct crimes and to describe the people, places, and things involved in them. Studies of homicides published in the forensic science literature show how trace evidence is used to solve crimes.

Examples and Cases

Trace evidence is found in many different forms with some examples including, but not limited, to fire debris, gunshot residue, glass fragments, and fibres. Each of these types of evidence will have a trained analyst in that specific field who will conduct the analysis on these items.

Some case examples of the use of trace evidence could include a Vehicular accident reconstruction which relies on marks such as tire impressions to estimate vehicle speed before and during an accident, as well as braking and impact forces. Fibres from clothing worn by a pedestrian found in the paint and/or road grime of a striking vehicle can be linked to the specific vehicle involved in a hit-and-run collision. "Witness marks" are also an important form of trace evidence, especially in engineering and may be critical in understanding how a product failed. A typical witness mark could be an impact depression which broke a product. It is especially useful if that mark can be linked to the product which made the impact such as a hammer or nail. Such marks are also commonly encountered in criminal cases and include bite marks, puncture marks, bullet holes, etc.

Recording and Recovery

After the trace evidence has been located at a crime scene the next step is to record it. There are multiple ways of recording evidence and that is dependant on the type of evidence found. The first option is to videotape the crime scene. This can be done to provide perspective on the actual location of the evidence throughout the entire scene. The next step in recording the evidence would be still photographs. The images included should be photos of the evidence both with flash and without, the evidence with a ruler for size reference, and the evidence with its number in the photo.

As for the recovery of the evidence, samples may be collected by handpicking, tape lifts, combing, or removal of an entire object. Hand picking is as straight forward as it sounds with the evidence being carefully picked up by hand or with forceps and placed in the proper packaging for that specific item. Tape lifts are used when the item cannot be picked up by hand or with forceps. It involves the sticky side of tape being placed on the evidence and then carefully lifting and placing it either on a sheet of clear acetate or on its original packaging. Combing is used when trace evidence needs to be removed from a person who is either alive or deceased. A different comb should be used for each piece of evidence being removed and the combs and paper must be packaged separate from the evidence. The removal of an entire object occurs when the evidence simply can not be removed from this object. The collection method used is entirely based on what the evidence type is and the surface it is on. It is also important to note that great care must be taken to prevent contamination of the evidence with other substances (such as natural oil and sweat on the hand of the collector).

Trace Evidence is also found in much smaller amounts at crime scenes. The Scientific Working Group Materials Analysis (SWGMAT) has created guidelines to ensure proper protection and collection of trace evidence. In this document you can find steps to ensure proper documentation, tips to avoid contamination and loss of evidence, proper detection, collection, and preservation techniques, as well as considerations for specific types of trace materials. The Federal Bureau of Investigation (FBI), has even implemented these standards into their work revolving trace evidence.

Following these standards and guidelines will ensure accurate analysis of crime scene evidence and increase the strength of the evidence in courts.

Analysis

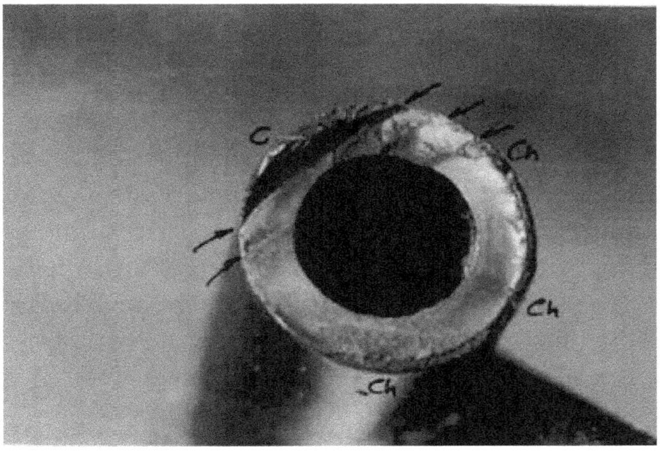

Fig 1: Close-Up of Broken Fuel Pipe Using Optical Microscopy

Fig 2: Droplets of Human Blood. The Droplets are Round and Show No Splattering, Indicating they Impacted Relatively Slowly, in this Case from a Height of Two Feet

The way the analysis of trace evidence is performed is dependant on the evidence being examined. Most commonly the trace evidence that is examined is microscopic due to the fact that it can not be examined by the naked eye. In this case there are many different types of microscopes that can be used including stereomicroscope, scanning electron microscope (SEM) or comparison microscope. SEM is especially useful because X-ray analysis can be conducted on selected areas of the sample. This can be especially useful as chemical residues can show unusual elements present which may indicate a chemical attack of the product. A car accident caused by a diesel fuel leak, for example, showed traces of sulfur on the cracked tube indicative of an attack by sulfuric acid from the battery.

When it comes to impression trace analysis such as a tire track in mud or a footprint, a cast or lift may be made of the impression. This would allow the analyst to identify what made the impression for example, the type of tire, the manufaturure as well as the condition of the tire.

The first step in examining gunshot residue is of course using a microscope to observe propellant residue particles which could include gun powder, lead and other materials depending on the type of propellant used. Tests like the Modified Griess Test and the Sodium Rhodizinate Test are wet chemical techniques used to develop residue that can not be seen. Small amounts of explosives, volatile hydrocarbons, and other chemicals are identified with the use of analytical instruments, such as gas chromatography, mass spectrometry, and infrared spectroscopy, all of which separate out the components of the chemicals.

Similar comments apply to damaged items from an accident scene, but care is needed in ensuring that the sample is not damaged by the testing or sampling for testing. Such nondestructive testing must always be used first before considering destructive methods which involve taking small samples from the item for more detailed tests, such as spectroscopic analysis. Use of all such methods must be done in consultation with other experts and the relevant authorities, such as lawyers on both sides of a case.

Problems

False positives and contamination by subsequent handling or nearby objects (e.g. mixing of blood from victim and attacker), are problems of many common substances and the necessity of human involvement in the collection of trace evidence. Both can occur with DNA traces and fingerprints, and therefore evidence should be collected, analyzed, and presented in accordance with established guidelines. Partial fingerprints are even more vulnerable to false positives. Samples from accidents or crimes should therefore be protected as much as possible by enclosure in a sealable container as soon as possible, after an incident is under investigation.

In terms of scientific research that is being conducted for trace evidence, there are many gaps and a lot that remains to be done. No standards and methodologies have been created to determine physical fit between two piece of evidence and consistency of work conducted from various analysts. Therefore, it is up to the analysts discretion to determine how likely it is that the two samples came from the same source. The Organization of Scientific Area Committees (OSAC) have clarified that future research must be conducted to create methods for quality of fit and error rates. Improving this field of forensic science will improve the quality of samples and develop quality case reports

1.3.6 Forensic Firearm Examination

Forensic firearm examination is the forensic process of examining the characteristics of firearms or bullets left behind at a crime scene. Specialists in this field try to link bullets to weapons and weapons to individuals. They can raise and record obliterated serial numbers in an attempt to find the registered owner of a weapon and look for fingerprints on a weapon and cartridges.

By examining unique striations impressed into a bullet from the barrel of a gun, expended ammunition can be linked back to a specific weapon. These striations are due to the rifling inside the barrels of firearms. Rifling spins the bullet when it is fired out of the barrel to improve precision. Although bullet striations are individualized unique evidence, microscopic striations in the barrel of the weapon are subject to change slightly, after each round that is fired. For this reason, forensic ballistics examiners may not fire more than five shots from a weapon found at a scene. Known exemplars taken from a seized weapon can be compared to samples recovered from a scene using a comparison microscope as well as newer 3-D imaging technology. Striation images can also be uploaded to national databases. Furthermore, the markings can be compared to other images in an attempt to link one weapon to multiple crime scenes.

Like all forensic specialties, forensic firearm examiners are subject to being called to testify in court as expert witnesses. However, the reliability of some techniques of forensic firearm examination have been criticized.

History

The ability to compare ammunition is a direct result of the invention of rifling around the turn of the 16th century. By forcing the bullet to spin as it travels down the barrel of the weapon the precision is greatly increased. At the same time, the rifling leaves marks on the bullet that are indicative of that particular barrel. Prior to mass production of firearms, each barrel and bullet mold was hand made by gunsmiths making them unique. The first successful documented case of forensic firearm examination occurred in 1835 when a member of the Bow Street Runners in London matched a recovered bullet from a murder victim to a specific mold in a suspect's home confirming that he made the bullet; this gave further evidence that the bullet maker was the perpetrator and he was convicted. As manufacturing and automation replaced hand tools, the ability to compare bullets became impossible due to the standardization of molds within a specific company. However, experts in the field postulated that there were microscopic differences on each barrel left during the manufacturing process. These differences were a result of wear on the machines and since each new weapon caused a tiny amount of wear, each barrel would be slightly different from every other barrel produced by that company. Also, each bullet fired from a specific barrel would be printed with the same marks, allowing investigators to identify the weapon that fired a specific bullet.

One of the first uses of this knowledge was in 1915 to exonerate Charles Stielow of the murder of his neighbors. Stielow was sentenced to death and appealed to Charles S. Whitman, the Governor of New York, who was not convinced by the evidence used to convict Stielow. Whitman halted the execution until an inquiry could be conducted and after further examination it was shown that Stielow's firearm could not have fired the bullets recovered from the victims. The invention of the comparison microscope by Calvin Goddard and Phillip O. Gravelle in 1925 modernized the forensic examination of firearms. Simultaneous comparison of two different objects at the same time allowed to closely examine striations for matches and therefore make a more definitive statement as to whether or not they matched.

One of the first true tests of this new technology was in the aftermath of the Saint Valentine's Day Massacre in 1929. During the Prohibition Era, competing gang members were fighting over bootlegging operations within the city of Chicago. Members of the Chicago Outfit and the Egan's Rats led by Al Capone attempted to remove all competition from Chicago by eliminating the North Side Gang leader Bugs Moran. The massacre missed Moran, who was not present, but killed seven members of the North Side Gang.

The murderers attempted to cover up their crime by posing as police officers, even dressing in police uniforms. Witnesses saw two "officers" leaving the scene, which implicated the Chicago police department as the perpetrators of the massacre. High levels of police corruption during that time period made it seem likely that the police department committed the killings. The investigation stalled until December 1929 when Fred Burke, a member of the Egan's Rats, shot and killed a police officer in St. Joseph, Michigan. Officers searching for Burke were led to a home in nearby Stevensville. While Burke was not there, inside officers found an arsenal of weapons including two Thompson submachine guns. The Chicago police department was contacted and the weapons were brought back to Chicago for testing. Goddard was asked to compare the weapons to collected evidence found at the massacre using his new "ballistic-forensics" technique. After test firing the guns, Goddard proved that the weapons were those used to kill the members of the

North Side Gang, absolving the Chicago police department of all involvement. The successful use of Goddard's technique resulted in the solidification of his place as the father of forensic firearm examination.

Examination of the Firearm

Any firearm collected during the course of an investigation could yield viable evidence if examined. For forensic firearm examination specific evidence that can be recovered include weapon serial numbers and potentially fingerprints left on the weapon's surface.

Fingerprint Recovery

Fingerprint recovery from the surface of firearms is done with cyanoacrylate (more commonly known as superglue) fuming. Firearms are placed in a specially designed fume hood designed to evenly distribute fumes instead of removing them. Liquid superglue is placed in a container and heated until it is in a gaseous state. The circulating fumes adhere to the oils left behind by the fingerprint, turning the print white. The resulting white print can be enhanced with fingerprint powder to increase the contrast of the white print against the weapon's finish. While using the fuming technique on recovered guns is commonplace, the recovery of fingerprints from the surfaces of a firearm is challenging due to the textured grip and the general condition of recovered weapons. If fingerprints are recovered, they can be processed through fingerprint databases such as the Integrated Automated Fingerprint Identification System (IAFIS). Various parts of the recovered weapon can also be tested for touch DNA left by whoever handled it. However, the low levels of DNA that can be recovered presents numerous issues such as contamination and analysis anomalies such as allele drop-out and drop-in.

Serial Number Recovery

Serial number examinations are a form of tool mark analysis. Serial numbers became commonplace after the United States passed the Gun Control Act of 1968. This law mandated that all guns manufactured in or imported into the country have a serial number. Prior to 1968, many firearms either did not have a serial number or the serial numbers were not unique and were reused by a manufacturer on multiple firearms. If a recovered weapon has had the serial numbers altered or destroyed, examiners can attempt to recover the original numbers. The two main methods for the restoration of serial numbers are magnetic particle inspection and chemical restoration. It is recommended that magnetic particle inspection be performed first due to the nondestructive nature of the method. If magnetic particle inspection fails, chemical restoration is the next step in the forensic analysis.

If the serial number is successfully restored it can be used to help investigators track the weapon's history, as well as potentially determine who owns the weapon.

Firearm databases such as the National Crime Information Center of the United States and INTERPOL's Firearm Reference Table can be used by investigators to track weapons that have been lost, stolen, or used previously in other crimes.

- **Magnetic Particle Inspection:** Originally developed as a method to detect flaws or irregularities in magnetic materials, magnetic particle inspection can be used on firearms to visualize the serial number underneath the obliterated area. When performing this technique, examiners place the weapon in a magnetic field. The irregularities in the metal, in this case the serial number, cause the field to deform. When a solution of ferrous particles is added to the weapon's magnetized surface they will be attracted to the area where the magnetic field has deformed and will build up in the area. If fluorescent particles are added to the ferrous solution, ultraviolet light can be used to make it easier to visualize any recovered serial number.
- **Chemical Restoration:** Chemical restoration is a type of chemical milling. Typically, chemical milling is used to slowly remove material to create a desired shape. In serial number restoration, small amounts of metal are removed until variations in the metal corresponding to the serial number are visible. This is possible because stamping the numbers distorts the grain boundary structure underneath the surface of the metal. However, chemical restoration is limited to that depth and is only successful when the obliteration of the serial number is superficial. Examiners performing a restoration first sand the area where the serial number used to be. This removes any debris from the area left when the serial number was obliterated. The examiner then chooses a chemical, usually an acid, that will be used to slowly bring the number back to the surface. The type of chemical that is used depends on the material the weapon is made of. These acids can range from Fry's Reagent for a magnetic metal, which is a mixture of hydrochloric acid, cupric chloride, and distilled water, to an acidic ferric chloride solution for a non-magnetic, non-aluminum material.

Examination of Cartridges

Spent cartridges found at a scene can be examined for physical evidence such as fingerprints or compared to samples that match them to a weapon. The examination of the cartridge relies on the unique tool marks left by the various parts of the weapon including the firing pin and the ejector in semi and fully automatic firearms. These markings can be compared and matched to known exemplars fired from the same weapon using the same parts. The examination of the marks left on the cartridge is done using a comparison microscope. Examiners view the questioned cartridge and the known exemplar simultaneously, looking for similar microscopic marks left during the firing process.

Cartridges are also routinely examined for fingerprints as the act of loading the ammunition into the magazine, or chamber, leaves recoverable impressions. These fingerprints can survive the firing processes and, while a rare occurrence, fingerprints have been obtained from cartridges recovered from the scene. Cartridges are subjected to cyanoacrylate fuming and examined for any usable prints. Usable prints are photographed and can be uploaded to fingerprint databases such as IAFIS for comparison with known exemplars. Cartridges can also be swabbed for trace DNA left by the individual who loaded the magazine. The extremely low levels of recoverable DNA present the same issues as swabbing a firearm for DNA.

Advancements in microscopic stamping have led to a push for the inclusion of firing pin microstamping. The microstamp is etched onto the firing pin and is transferred to the cartridge during the firing process. Each firing pin would have a unique serial number allowing investigators to trace casings found at a crime scene to a known firearm. The practice is not in use as of 2024, although California has enacted legislation that requires microstamping on all newly sold firearms. The law, and microstamping in general, has received significant opposition from gun manufacturers due the technology being unreliable, and not proven to aid in preventing or solving crimes.

Examination of Bullets

Preliminary examination of the bullet can exclude a large number of weapons by examining the general characteristics of a recovered bullet. By determining general aspects of the fired ammunition, a number of weapons can be immediately excluded as being incapable of firing that type of bullet. The make and model of the weapon can also be inferred from the combination of different class characteristics that are common to specific manufactures.

The three main class characteristics of all bullets are the lands and grooves, the caliber of the bullet, and the rifling twist. All three can be tied directly to the type of barrel that was used to fire the bullet. The lands and grooves of barrel are the bumps and valleys created when the rifling is created. The caliber is the diameter of the barrel. The twist is the direction of the striations left by the barrel's rifling, clockwise (right-handed) or counterclockwise (left-handed). Most barrels will have a right-handed twist with the exception of weapons created by the Colt's Manufacturing Company which uses left-handed twists. Weapon barrels that match the class characteristics of recovered bullets can be examined further for individual characteristics to determine if the bullet came from that particular weapon.

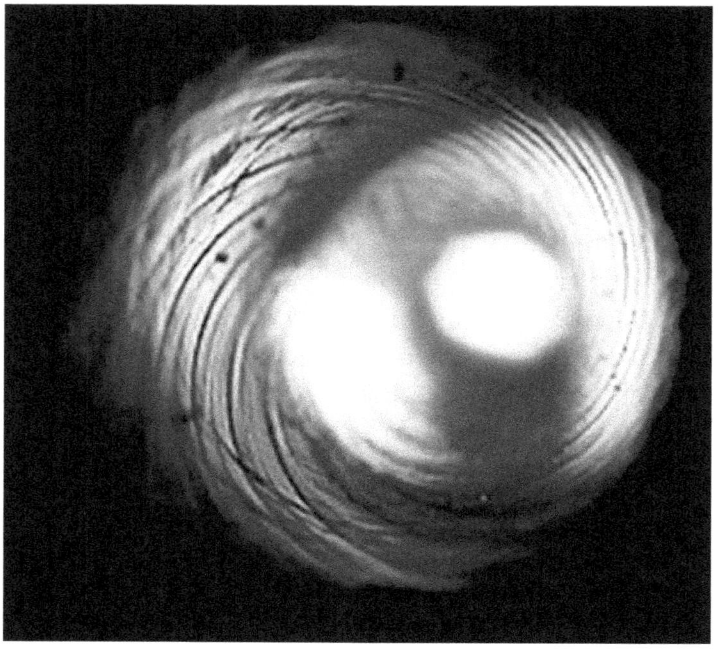

Fig 3: Rifling Pattern for a Remington Rifle Showing a Clockwise (Right-Handed) Twist

Class Characteristics

Individual Characteristics

In order to compare individual striations, examiners must obtain a known sample using the seized weapon.

For slower-traveling bullets, such as pistols or revolvers, known bullet exemplars are created by firing the weapon into a water tank. The spent bullet can be recovered, intact, as the water slows down the bullet before it can reach the tank walls. For faster traveling bullets, such as those fired from high-powered rifles and military style weapons, water tanks cannot be used as the tank will not provide enough stopping power for the projectiles. To examine these weapons, investigators must fire them at a target at a controlled range with enough backing to stop the bullet and collect the spent round after it has been fired.

Once a known exemplar is produced, the evidence sample can be compared to the known by examining both at the same time with a comparison microscope. Striations that line up are examined more closely, looking for multiple consecutive matches. There is no set number of consecutive matches that equates to a match declaration, and examiners are trained to use the phrase "sufficient agreement" when testifying. The degree to which an examiner can make that determination is based on their training and expertise. All findings by examiners are subject to questioning by both sides, prosecution and defense, during testimony in court.

Striation Databasing

Bullets and casings found at a scene require a known example to compare to in order to match them to a weapon. Without a weapon, the striation pattern can be uploaded to a database such as the National Integrated Ballistic Identification Network (NIBIN) maintained by the ATF or the United Kingdom's National Ballistics Intelligence Service (NABIS). Information uploaded to these databases can be used to track gun crimes and to link crimes together. Maintainers of these databases recommend that every recovered firearm be test fired and the resulting known exemplar be uploaded into the database.

In the 1990s, there were two databases that were formed for storage of pictures of shell casings and bullets in gun crimes. The first was the Drugfire system which was used by the FBI. The second, the IBIS (Integrated Ballistic Identification System) was created by Forensic Technology, Inc. and eventually bought by the Alcohol Tobacco and Firearms (ATF) in 1993. The FBI and ATF realized that their systems would not work together, and they needed to find a way to share information between them. The NIBIN board was created in 1997, in hopes of creating one imaging system. A year after the creation of the NIBIN board, both the ATF and FBI decided to put their resources together toward one of the systems, and created the National Integrated Ballistics Information Network, with IBIS as the system.

Criticisms

Firearm examiners have attempted to determine the shooter's position by the location of spent bullet casings. The use of ejection pattern studies were originally part of incident reconstruction and methods for determining shooter location continue to be explained in major crime scene examination books. However, the validity of ejection pattern analysis has been brought into question by multiple studies that look at the reproducibility and end determination of shooter position by qualified examiners. Studies have shown that over 25% of spent casings land somewhere other than to the right and rear of the shooter. This is the most commonly accepted location for where spent cartridge casings should fall, and the large percentage of casings that end up somewhere else raises concerns for the validity of the examination technique. Investigators should only present a location gained from an ejection pattern study as a tentative estimate when using the information in a courtroom setting.

Prior to September 2005, comparative bullet-lead analysis was performed on bullets found at a scene that were too destroyed for striation comparison. The technique would attempt to determine the unique elemental breakdown of the bullet and compare it to seized bullets possessed by a suspect. Review of the method found that the breakdown of elements found in bullets could be significantly different enough to potentially allow for two bullets from separate sources to be correlated to each other. However, there are not enough differences to definitely match a bullet from a crime scene to one taken from a suspect's possession. An additional report in 2004 from the National Academy of Sciences (NAS) found that the testimony given regarding comparative bullet-lead analysis was overstated and potentially "misleading under the federal rules of evidence". In 2005, the Federal Bureau of Investigation indicated that they would no longer be performing comparative bullet-lead analysis.

Further criticism came from the 2009 NAS report on the current state of various forensic fields in the United States. The report's section on firearm examination focused on the lack of defined requirements that are necessary in order to determine "matches" between known and unknown striations. The NAS stated that, "sufficient studies have not been done to understand the reliability and repeatability of the methods." Without defined procedures on what is and what isn't considered "sufficient agreement" the report states that forensic firearm examination contains fundamental problems that need to be addressed by the forensic community through a set of repeatable scientific studies that outline standard operating procedures that should be adopted by all firearm examiners. Another report issued in 2016 by the United States President's Council of Advisors on Science and Technology confirmed the NAS's findings, finding only one appropriately designed study that examined the rate of false positives and reliability amongst firearm examiners.

In 2020 Itiel E. Dror and Nicholas Scurich looked at the validity of ballistic forensic experts when attempting to make an identification of a shell or bullet. They found that while some experts would come to the conclusion that the bullets were a definite match, another expert looking at the same evidence would determine it inconclusive. Dror and Scurich argue an "inconclusive" determination affects the error rate for the study, and provides very little confidence in the overall findings of the scientists. According to Dror and Scurich, the error rate, which was zero to one percent, could be higher.

Their reasoning behind this is that if an "answer" was marked as inconclusive, it must count as a correct answer which decreases the error rate making it lower than it probably should be.They wondered how different the error rate would be if inconclusive was not an option. In addition, Dror and Scruich noted that the scientists seemed to come up with a more conclusive decision on the evidence if there was the added part of a human life hanging in the balance. In 2021 Alex Biederman and Kyriakos N. Kotsoglou responded to the Dror and Scurich paper and raised issues. Some of the issues raised by Biederman and Kotsoglou included: a paradox in which examiners' results agreed with ground truth but would be considered "error" via Dror and Scurich's proposals. Biederman and Kotsoglou also pointed out that Dror and Scurich's proposals would set false incentives where examiners would be directed to be "divining what the mythical forensic wisdom of the consensus opinion might (and hence enshrine the false belief in the existence of such wisdom), be rather than the ground truth". Biederman and Kotsoglou concluded "In all, our analysis does not leave much intact from recent attempts to label 'inconclusives as errors."

1.4 ANALYTICAL TECHNIQUES IN FORENSIC CHEMISTRY

Forensic chemistry relies on a variety of sophisticated analytical techniques to examine physical evidence. These techniques provide the precision and accuracy necessary for reliable forensic investigations.

1.4.1 Spectroscopy

Spectroscopy is the field of study that measures and interprets electromagnetic spectra. In narrower contexts, spectroscopy is the precise study of color as generalized from visible light to all bands of the electromagnetic spectrum.

Spectroscopy, primarily in the electromagnetic spectrum, is a fundamental exploratory tool in the fields of astronomy, chemistry, materials science, and physics, allowing the composition, physical structure and electronic structure of matter to be investigated at the atomic, molecular and macro scale, and over astronomical distances.

Historically, spectroscopy originated as the study of the wavelength dependence of the absorption by gas phase matter of visible light dispersed by a prism. Current applications of spectroscopy include biomedical spectroscopy in the areas of tissue analysis and medical imaging. Matter waves and acoustic waves can also be considered forms of radiative energy, and recently gravitational waves have been associated with a spectral signature in the context of the Laser Interferometer Gravitational-Wave Observatory (LIGO).

Spectroscopy is a branch of science concerned with the spectra of electromagnetic radiation as a function of its wavelength or frequency measured by spectrographic equipment, and other techniques, in order to obtain information concerning the structure and properties of matter.

Spectral measurement devices are referred to as spectrometers, spectrophotometers, spectrographs or spectral analyzers. Most spectroscopic analysis in the laboratory starts with a sample to be analyzed, then a light source is chosen from any desired range of the light spectrum, then the light goes through the sample to a dispersion array (diffraction grating instrument) and captured by a photodiode. For astronomical purposes, the telescope must be equipped with the light dispersion device. There are various versions of this basic setup that may be employed.

Spectroscopy began with Isaac Newton splitting light with a prism; a key moment in the development of modern optics. Therefore, it was originally the study of visible light that we call color that later under the studies of James Clerk Maxwell came to include the entire electromagnetic spectrum. Although color is involved in spectroscopy, it is not equated with the color of elements or objects that involve the absorption and reflection of certain electromagnetic waves to give objects a sense of color to our eyes. Rather spectroscopy involves the splitting of light by a prism, diffraction grating, or similar instrument, to give off a particular discrete line pattern called a "spectrum" unique to each different type of element. Most elements are first put into a gaseous phase to allow the spectra to be examined although today other methods can be used on different phases. Each element that is diffracted by a prism-like instrument displays either an absorption spectrum or an emission spectrum depending upon whether the element is being cooled or heated.

Until recently all spectroscopy involved the study of line spectra and most spectroscopy still does. Vibrational spectroscopy is the branch of spectroscopy that studies the spectra.

However, the latest developments in spectroscopy can sometimes dispense with the dispersion technique. In biochemical spectroscopy, information can be gathered about biological tissue by absorption and light scattering techniques. Light scattering spectroscopy is a type of reflectance spectroscopy that determines tissue structures by examining elastic scattering. In such a case, it is the tissue that acts as a diffraction or dispersion mechanism.

Spectroscopic studies were central to the development of quantum mechanics, because the first useful atomic models described the spectra of hydrogen, which include the Bohr model, the Schrödinger equation, and Matrix mechanics, all of which can produce the spectral lines of hydrogen, therefore providing the basis for discrete quantum jumps to match the discrete hydrogen spectrum. Also, Max Planck's explanation of blackbody radiation involved spectroscopy because he was comparing the wavelength of light using a photometer to the temperature of a Black Body. Spectroscopy is used in physical and analytical chemistry because atoms and molecules have unique spectra. As a result, these spectra can be used to detect, identify and quantify information about the atoms and molecules. Spectroscopy is also used in astronomy and remote sensing on Earth. Most research telescopes have spectrographs. The measured spectra are used to determine the chemical composition and physical properties of astronomical objects (such as their temperature, density of elements in a star, velocity, black holes and more). An important use for spectroscopy is in biochemistry. Molecular samples may be analyzed for species identification and energy content.

History

The history of spectroscopy began with Isaac Newton's optics experiments (1666–1672). According to Andrew Fraknoi and David Morrison, "In 1672, in the first paper that he submitted to the Royal Society, Isaac Newton described an experiment in which he permitted sunlight to pass through a small hole and then through a prism. Newton found that sunlight, which looks white to us, is actually made up of a mixture of all the colors of the rainbow." Newton applied the word "spectrum" to describe the rainbow of colors that combine to form white light and that are revealed when the white light is passed through a prism.

Fraknoi and Morrison state that "In 1802, William Hyde Wollaston built an improved spectrometer that included a lens to focus the Sun's spectrum on a screen. Upon use, Wollaston realized that the colors were not spread uniformly, but instead had missing patches of colors, which appeared as dark bands in the spectrum." During the early 1800s, Joseph von Fraunhofer made experimental advances with dispersive spectrometers that enabled spectroscopy to become a more precise and quantitative scientific technique. Since then, spectroscopy has played and continues to play a significant role in chemistry, physics, and astronomy. Per Fraknoi and Morrison, "Later, in 1815, German physicist Joseph Fraunhofer also examined the solar spectrum, and found about 600 such dark lines (missing colors), are now known as Fraunhofer lines, or Absorption lines."

In quantum mechanical systems, the analogous resonance is a coupling of two quantum mechanical stationary states of one system, such as an atom, via an oscillatory source of energy such as a photon. The coupling of the two states is strongest when the energy of the source matches the energy difference between the two states.

The energy E of a photon is related to its frequency ν by $E = h\nu$ where h is the Planck constant, and so a spectrum of the system response vs. photon frequency will peak at the resonant frequency or energy. Particles such as electrons and neutrons have a comparable relationship, the de Broglie relations, between their kinetic energy and their wavelength and frequency and therefore can also excite resonant interactions.

Spectra of atoms and molecules often consist of a series of spectral lines, each one representing a resonance between two different quantum states. The explanation of these series, and the spectral patterns associated with them, were one of the experimental enigmas that drove the development and acceptance of quantum mechanics. The hydrogen spectral series in particular was first successfully explained by the Rutherford–Bohr quantum model of the hydrogen atom. In some cases spectral lines are well separated and distinguishable, but spectral lines can also overlap and appear to be a single transition if the density of energy states is high enough. Named series of lines include the principal, sharp, diffuse and fundamental series.

Theory

The underlying premise of spectroscopy is that light is made of different wavelengths and that each wavelength corresponds to a different frequency. The importance of spectroscopy is centered around the fact that every element in the periodic table has a unique light spectrum described by the frequencies of light it emits or absorbs consistently appearing in the same part of the electromagnetic spectrum when that light is diffracted.

This opened up an entire field of study with anything that contains atoms. Spectroscopy is the key to understanding the atomic properties of all matter.

As such spectroscopy opened up many new sub-fields of science yet undiscovered. The idea that each atomic element has its unique spectral signature enabled spectroscopy to be used in a broad number of fields each with a specific goal achieved by different spectroscopic procedures. The National Institute of Standards and Technology maintains a public Atomic Spectra Database that is continually updated with precise measurements.

The broadening of the field of spectroscopy is due to the fact that any part of the electromagnetic spectrum may be used to analyze a sample from the infrared to the ultraviolet telling scientists different properties about the very same sample.

For instance in chemical analysis, the most common types of spectroscopy include atomic spectroscopy, infrared spectroscopy, ultraviolet and visible spectroscopy, Raman spectroscopy and nuclear magnetic resonance.

In nuclear magnetic resonance (NMR), the theory behind it is that frequency is analogous to resonance and its corresponding resonant frequency. Resonances by the frequency were first characterized in mechanical systems such as pendulums, which have a frequency of motion noted famously by Galileo.

Classification of Methods

Spectroscopy is a sufficiently broad field that many sub-disciplines exist, each with numerous implementations of specific spectroscopic techniques. The various implementations and techniques can be classified in several ways.

Type of Radiative Energy

The types of spectroscopy are distinguished by the type of radiative energy involved in the interaction. In many applications, the spectrum is determined by measuring changes in the intensity or frequency of this energy. The types of radiative energy studied include:

- Electromagnetic radiation was the first source of energy used for spectroscopic studies. Techniques that employ electromagnetic radiation are typically classified by the wavelength region of the spectrum and include microwave, terahertz, infrared, near-infrared, ultraviolet-visible, x-ray, and gamma spectroscopy.
- Particles, because of their de Broglie waves, can also be a source of radiative energy. Both electron and neutron spectroscopy are commonly used. For a particle, its kinetic energy determines its wavelength.
- Acoustic spectroscopy involves radiated pressure waves.
- Dynamic mechanical analysis can be employed to impart radiating energy, similar to acoustic waves, to solid materials.

Nature of the Interaction

The types of spectroscopy also can be distinguished by the nature of the interaction between the energy and the material. These interactions include:

- **Absorption Spectroscopy:** Absorption occurs when energy from the radiative source is absorbed by the material. Absorption is often determined by measuring the fraction of energy transmitted through the material, with absorption decreasing the transmitted portion.
- **Emission Spectroscopy:** Emission indicates that radiative energy is released by the material. A material's blackbody spectrum is a spontaneous emission spectrum determined by its temperature. This feature can be measured in the infrared by instruments such as the atmospheric emitted radiance interferometer. Emission can also be induced by other sources of energy such as flames, sparks, electric arcs or electromagnetic radiation in the case of fluorescence.
- **Elastic Scattering and Reflection Spectroscopy:** Determine how incident radiation is reflected or scattered by a material. Crystallography employs the scattering of high energy radiation, such as x-rays and electrons, to examine the arrangement of atoms in proteins and solid crystals.
- **Impedance Spectroscopy:** Impedance is the ability of a medium to impede or slow the transmittance of energy. For optical applications, this is characterized by the index of refraction.
- **Inelastic Scattering Phenomena:** Involve an exchange of energy between the radiation and the matter that shifts the wavelength of the scattered radiation. These include Raman and Compton scattering.
- Coherent or Resonance Spectroscopy are techniques where the radiative

energy couples two quantum states of the material in a coherent interaction that is sustained by the radiating field. The coherence can be disrupted by other interactions, such as particle collisions and energy transfer, and so often require high intensity radiation to be sustained. Nuclear magnetic resonance (NMR) spectroscopy is a widely used resonance method, and ultrafast laser spectroscopy is also possible in the infrared and visible spectral regions.

- Nuclear spectroscopy are methods that use the properties of specific nuclei to probe the local structure in matter, mainly condensed matter, molecules in liquids or frozen liquids and bio-molecules.
- Quantum logic spectroscopy is a general technique used in ion traps that enables precision spectroscopy of ions with internal structures that preclude laser cooling, state manipulation, and detection. Quantum logic operations enable a controllable ion to exchange information with a co-trapped ion that has a complex or unknown electronic structure.

Type of Material

Spectroscopic studies are designed so that the radiant energy interacts with specific types of matter.

- **Atoms:** Atomic spectroscopy was the first application of spectroscopy. Atomic absorption spectroscopy and atomic emission spectroscopy involve visible and ultraviolet light. These absorptions and emissions, often referred to as atomic spectral lines, are due to electronic transitions of outer shell electrons as they rise and fall from one electron orbit to another. Atoms also have distinct x-ray spectra that are attributable to the excitation of inner shell electrons to excited states. Atoms of different elements have distinct spectra and therefore atomic spectroscopy allows for the identification and quantitation of a sample's elemental composition. After inventing the spectroscope, Robert Bunsen and Gustav Kirchhoff discovered new elements by observing their emission spectra. Atomic absorption lines are observed in the solar spectrum and referred to as Fraunhofer lines after their discoverer. A comprehensive explanation of the hydrogen spectrum was an early success of quantum mechanics and explained the Lamb shift observed in the hydrogen spectrum, which further led to the development of quantum electrodynamics. Modern implementations of atomic spectroscopy for studying visible and ultraviolet transitions include flame emission spectroscopy, inductively coupled plasma atomic emission spectroscopy, glow discharge spectroscopy, microwave induced plasma spectroscopy, and spark or arc emission spectroscopy. Techniques for studying x-ray spectra include X-ray spectroscopy and X-ray fluorescence.

- **Molecules:** The combination of atoms into molecules leads to the creation of unique types of energetic states and therefore unique spectra of the transitions between these states. Molecular spectra can be obtained due to electron spin states (electron paramagnetic resonance), molecular rotations, molecular vibration, and electronic states. Rotations are collective motions of the atomic nuclei and typically lead to spectra in the microwave and millimetre-wave spectral regions. Rotational spectroscopy and microwave spectroscopy are synonymous. Vibrations are relative motions of the atomic nuclei and are studied by both infrared and Raman spectroscopy. Electronic excitations are studied using visible and ultraviolet spectroscopy as well as fluorescence spectroscopy. Studies in molecular spectroscopy led to the development of the first maser and contributed to the subsequent development of the laser.
- **Crystals and Extended Materials:** The combination of atoms or molecules into crystals or other extended forms leads to the creation of additional energetic states. These states are numerous and therefore have a high density of states. This high density often makes the spectra weaker and less distinct, i.e., broader. For instance, blackbody radiation is due to the thermal motions of atoms and molecules within a material. Acoustic and mechanical responses are due to collective motions as well. Pure crystals, though, can have distinct spectral transitions, and the crystal arrangement also has an effect on the observed molecular spectra. The regular lattice structure of crystals also scatters x-rays, electrons or neutrons allowing for crystallographic studies.
- **Nuclei:** Nuclei also have distinct energy states that are widely separated and lead to gamma ray spectra. Distinct nuclear spin states can have their energy separated by a magnetic field, and this allows for nuclear magnetic resonance spectroscopy.

Other Types

Other types of spectroscopy are distinguished by specific applications or implementations:

- Acoustic resonance spectroscopy is based on sound waves primarily in the audible and ultrasonic regions.
- Auger electron spectroscopy is a method used to study surfaces of

materials on a micro-scale. It is often used in connection with electron microscopy.

- Cavity ring-down spectroscopy.
- Circular dichroism spectroscopy.
- Coherent anti-Stokes Raman spectroscopy is a recent technique that has high sensitivity and powerful applications for *in vivo* spectroscopy and imaging.
- Cold vapour atomic fluorescence spectroscopy.
- Correlation spectroscopy encompasses several types of two-dimensional NMR spectroscopy.
- Deep-level transient spectroscopy measures concentration and analyzes parameters of electrically active defects in semiconducting materials.
- Dielectric spectroscopy.
- Dual-polarization interferometry measures the real and imaginary components of the complex refractive index.
- Electron energy loss spectroscopy in transmission electron microscopy.
- Electron phenomenological spectroscopy measures the physicochemical properties and characteristics of the electronic structure of multicomponent and complex molecular systems.
- Electron paramagnetic resonance spectroscopy.
- Force spectroscopy.
- Fourier-transform spectroscopy is an efficient method for processing spectra data obtained using interferometers. Fourier-transform infrared spectroscopy is a common implementation of infrared spectroscopy. NMR also employs Fourier transforms.
- Gamma spectroscopy.
- Hadron spectroscopy studies the energy/mass spectrum of hadrons according to spin, parity, and other particle properties. Baryon spectroscopy and meson spectroscopy are types of hadron spectroscopy.
- Multispectral imaging and hyperspectral imaging is a method to create a complete picture of the environment or various objects, each pixel containing a full visible, visible near infrared, near infrared, or infrared spectrum.
- Inelastic electron tunneling spectroscopy uses the changes in current due to inelastic electron-vibration interaction at specific energies that can also measure optically forbidden transitions.

- Inelastic neutron scattering is similar to Raman spectroscopy, but uses neutrons instead of photons.
- Laser-induced breakdown spectroscopy, also called laser-induced plasma spectrometry.
- Laser spectroscopy uses tunable lasers and other types of coherent emission sources, such as optical parametric oscillators, for selective excitation of atomic or molecular species.
- Light scattering spectroscopy (LSS) is a spectroscopic technique typically used to evaluate morphological changes in epithelial cells in order to study mucosal tissue and detect early cancer and precancer.
- Mass spectroscopy is a historical term used to refer to mass spectrometry. The current recommendation is to use the latter term. The term "mass spectroscopy" originated in the use of phosphor screens to detect ions.
- Mössbauer spectroscopy probes the properties of specific isotopic nuclei in different atomic environments by analyzing the resonant absorption of gamma rays.
- Multivariate optical computing is an all optical compressed sensing technique, generally used in harsh environments, that directly calculates chemical information from a spectrum as analogue output.
- Neutron spin echo spectroscopy measures internal dynamics in proteins and other soft matter systems.
- Nuclear quadrupole resonance is a chemical spectroscopy method mediated by NMR of the electric field gradient (EFG) in the absence of magnetic field.
- Perturbed angular correlation (PAC) uses radioactive nuclei as probe to study electric and magnetic fields (hyperfine interactions) in crystals (condensed matter) and bio-molecules.
- Photoacoustic spectroscopy measures the sound waves produced upon the absorption of radiation.
- Photoemission spectroscopy.
- Photothermal spectroscopy measures heat evolved upon absorption of radiation.
- Pump-probe spectroscopy can use ultrafast laser pulses to measure reaction intermediates in the femtosecond timescale.
- Raman optical activity spectroscopy exploits Raman scattering and

optical activity effects to reveal detailed information on chiral centers in molecules.

- Raman spectroscopy.
- Saturated spectroscopy.
- Scanning tunneling spectroscopy.
- Spectrophotometry.
- Spin noise spectroscopy traces spontaneous fluctuations of electronic and nuclear spins.
- Time-resolved spectroscopy measures the decay rates of excited states using various spectroscopic methods.
- Time-stretch spectroscopy.
- Thermal infrared spectroscopy measures thermal radiation emitted from materials and surfaces and is used to determine the type of bonds present in a sample as well as their lattice environment. The techniques are widely used by organic chemists, mineralogists, and planetary scientists.
- Transient grating spectroscopy measures quasiparticle propagation. It can track changes in metallic materials as they are irradiated.
- Ultraviolet photoelectron spectroscopy.
- Ultraviolet–visible spectroscopy.
- Vibrational circular dichroism spectroscopy.
- Video spectroscopy.
- X-ray photoelectron spectroscopy.

Applications

There are several applications of spectroscopy in the fields of medicine, physics, chemistry, and astronomy. Taking advantage of the properties of absorbance and with astronomy emission, spectroscopy can be used to identify certain states of nature. The uses of spectroscopy in so many different fields and for so many different applications has caused specialty scientific subfields. Such examples include:

- Determining the atomic structure of a sample.
- Studying spectral emission lines of the sun and distant galaxies.
- Space exploration.
- Cure monitoring of composites using optical fibers.

- Estimating weathered wood exposure times using near infrared spectroscopy.
- Measurement of different compounds in food samples by absorption spectroscopy both in visible and infrared spectrum.
- Measurement of toxic compounds in blood samples.
- Non-destructive elemental analysis by X-ray fluorescence.
- Electronic structure research with various spectroscopes.
- Redshift to determine the speed and velocity of a distant object.
- Determining the metabolic structure of a muscle.
- Monitoring dissolved oxygen content in freshwater and marine ecosystems.
- Altering the structure of drugs to improve effectiveness.
- Characterization of proteins.
- Respiratory gas analysis in hospitals.
- Finding the physical properties of a distant star or nearby exoplanet using the Relativistic Doppler effect.
- In-ovo sexing: spectroscopy allows to determine the sex of the egg while it is hatching. Developed by French and German companies, both countries decided to ban chick culling, mostly done through a macerator, in 2022.
- Process monitoring in Industrial process control.

1.4.2 Chromatography

Chromatography is essential in forensic chemistry for separating complex mixtures of chemicals. Gas chromatography (GC) and liquid chromatography (LC) are commonly employed to analyze drugs, toxic substances, and explosive residues.

In conjunction with mass spectrometry (MS), these techniques can provide detailed chemical profiles that help identify unknown substances.

1.4.3 Microscopy

Various forms of microscopy, including light microscopy and scanning electron microscopy (SEM), allow forensic chemists to examine physical evidence at the micro level.

Microscopic analysis is crucial for studying trace evidence like hair, fibers, and gunshot residues, as it enables forensic scientists to identify materials based on their physical and chemical characteristics.

1.4.4 Mass Spectrometry

Mass spectrometry (MS) is a powerful analytical tool used to identify chemical compounds by measuring the mass-to-charge ratio of ions. When coupled with gas chromatography (GC-MS) or liquid chromatography (LC-MS), it allows for the identification of specific chemicals in complex mixtures, such as drugs in biological samples or explosive residues in fire debris.

1.4.5 Electrophoresis

Electrophoresis is another technique used in forensic chemistry, particularly for the separation of DNA, proteins, and other biomolecules. Capillary electrophoresis (CE) is frequently used in DNA profiling, which plays a critical role in identifying individuals in criminal cases.

1.5 CHALLENGES AND LIMITATIONS IN FORENSIC CHEMISTRY

Despite its importance, forensic chemistry faces several challenges. The accuracy of the analyses depends on the integrity of the evidence collected and the proper maintenance of the chain of custody. Contamination, improper storage, or delays in testing can compromise results. Moreover, forensic chemists often work under pressure, as their findings are frequently needed to make swift legal decisions.

Another limitation is the potential for human error, whether in interpreting data or presenting findings in court. The misuse or misrepresentation of forensic evidence has led to wrongful convictions in the past. Thus, ongoing developments in technology and methodology are crucial to minimizing errors and enhancing the reliability of forensic chemical analyses.

1.6 CONCLUSION

Forensic chemistry, combined with the analysis of physical evidence, forms the foundation of modern criminal investigations. By utilizing advanced techniques in spectroscopy, chromatography, microscopy, and mass spectrometry, forensic chemists provide critical information that helps solve crimes, bring criminals to justice, and exonerate the innocent. As technology continues to advance, forensic chemistry will play an even more integral role in ensuring the integrity of the legal process. Understanding the complexities of forensic chemistry and its relationship with physical evidence is key to appreciating the broader field of forensic science.

CHAPTER 2
CHEMICAL TESTS

2.1 UNDERSTANDING THE CHEMICAL TESTS

Forensic chemistry plays a critical role in modern criminal investigations, providing methods for the detection, analysis, and identification of substances and materials found at crime scenes. A major component of this field involves chemical tests, which are used to examine various forms of physical evidence such as drugs, explosives, poisons, and biological materials. These tests help forensic scientists determine the nature of the substances involved, their origin, and how they may relate to criminal activity. The accuracy, reliability, and specificity of these chemical tests make them indispensable in the pursuit of justice.

2.1.1 Purposes

Chemical testing might have a variety of purposes, such as to:

- Determine if, or verify that, the requirements of a specification, regulation, or contract are met.
- Decide if a new product development program is on track: Demonstrate proof of concept.
- Demonstrate the utility of a proposed patent.
- Determine the interactions of a sample with other known substances.
- Determine the composition of a sample.
- Provide standard data for other scientific, medical, and Quality assurance functions.
- Validate suitability for end-use.
- Provide a basis for Technical communication.
- Provide a technical means of comparison of several options.
- Provide evidence in legal proceedings.

2.1.2 Biochemical Tests

- Clinistrips quantitatively test for sugar in urine.
- The Kastle-Meyer test tests for the presence of hemoglobin.

- Salicylate testing is a category of drug testing that is focused on detecting salicylates such as acetylsalicylic acid for either biochemical or medical purposes.
- The Phadebas test tests for the presence of saliva for forensic purposes.
- Iodine solution tests for starch.
- The Van Slyke determination tests for specific amino acids.
- The Zimmermann test tests for ketosteroids.
- Seliwanoff's test differentiates between aldose and ketose sugars.
- Test for lipids: add ethanol to sample, then shake; add water to the solution, and shake again. If fat is present, the product turns milky white.
- The Sakaguchi test detects the presence of arginine in protein.
- The Hopkins–Cole reaction tests for the presence of tryptophan in proteins.
- The nitroprusside reaction tests for the presence of free thiol groups of cysteine in proteins.
- The Sullivan reaction tests for the presence of cysteine and cystine in proteins.
- The Acree–Rosenheim reaction tests for the presence of tryptophan in proteins.
- The Pauly reaction tests for the presence of tyrosine or histidine in proteins.
- Heller's test tests for the presence of albumin in urine.
- Gmelin's test tests for the presence of bile pigments in urine.
- Hay's test tests for the presence of bile pigments in urine.

Reducing Sugars

- Barfoed's test tests for reducing polysaccharides or disaccharides.
- Benedict's reagent tests for reducing sugars or aldehydes.
- Fehling's solution tests for reducing sugars or aldehydes, similar to Benedict's reagent.
- Molisch's test tests for carbohydrates.
- Nylander's test tests for reducing sugars.
- Rapid furfural test distinguishes between glucose and fructose.

Proteins and Polypeptides

- The bicinchoninic acid assay tests for proteins.
- The Biuret test tests for proteins and polypeptides.
- Bradford protein assay measures protein quantitatively.
- The Phadebas amylase test determines alpha-amylase activity.

2.1.3 Organic Tests

- The carbylamine reaction tests for primary amines.
- The esterification reaction tests for the presence of alcohol and/or carboxylic acids.
- The Griess test tests for organic nitrite compounds.
- The 2,4-dinitrophenylhydrazine tests for carbonyl compounds.
- The iodoform reaction tests for the presence of methyl ketones, or compounds which can be oxidized to methyl ketones.
- The Schiff test detects aldehydes.
- Tollens' reagent tests for aldehydes (known as the silver mirror test).
- The Zeisel determination tests for the presence of esters or ethers.
- Lucas' reagent is used to distinguish between primary, secondary and tertiary alcohols.
- The bromine test is used to test for the presence of unsaturation and phenols.

2.1.4 Inorganic Tests

- Barium chloride tests for sulfates.
- Acidified silver nitrate solution tests for halide ions.
- The Beilstein test tests for halides qualitatively.
- The bead test tests for certain metals.
- The Carius halogen method measures halides quantitatively.
- Chemical tests for cyanide test for the presence of cyanide, CN^-.
- Copper sulfate tests for the presence of water.
- Flame tests test for metals.
- The Gilman test tests for the presence of a Grignard reagent.
- The Kjeldahl method quantitatively determines the presence of nitrogen.

- Nessler's reagent tests for the presence of ammonia.
- Ninhydrin tests for ammonia or primary amines.
- Phosphate tests test for phosphate.
- The sodium fusion test tests for the presence of nitrogen, sulfur, and halides in a sample.
- The Zerewitinoff determination tests for any acidic hydrogen.
- The Oddy test tests for acid, aldehydes, and sulfides.
- Gunzberg's test tests for the presence of hydrochloric acid.
- Kelling's test tests for the presence of lactic acid.

2.2 THE IMPORTANCE OF CHEMICAL TESTS IN FORENSIC SCIENCE

Chemical tests are fundamental to the field of forensic chemistry as they allow for the precise identification and analysis of substances that may be linked to criminal acts. These tests are performed on various materials, including drugs, poisons, bodily fluids, and trace evidence, and often serve as critical pieces of evidence in court. Through these tests, forensic scientists can identify illicit drugs in a suspect's possession, confirm the presence of toxic substances in a victim's body, or detect explosive residues at a crime scene.

In addition to identification, chemical tests can determine the quantity and purity of substances, which is crucial for understanding the context of the crime. For example, the amount of a drug found at a crime scene can indicate whether it was intended for personal use or distribution, impacting the severity of charges. Furthermore, chemical tests help reconstruct the events leading up to a crime by analyzing materials such as paint, glass, and fibers to establish connections between suspects, victims, and locations.

2.3 TYPES OF CHEMICAL TESTS IN FORENSIC CHEMISTRY

A variety of chemical tests are used in forensic investigations, each with its own specific applications and advantages. These tests can be broadly categorized into presumptive and confirmatory tests, with each playing a unique role in the analysis of forensic evidence.

2.3.1 Presumptive Tests

Presumptive tests are quick, preliminary tests used to indicate the possible presence of a substance. While they provide rapid results, they are not definitive and must be followed by more precise confirmatory tests. However, they are highly valuable in guiding investigators toward the next steps in evidence analysis.

Color Tests

Color tests are simple chemical tests that rely on the reaction of a substance with a reagent to produce a color change, indicating the presence of a specific chemical compound. These tests are commonly used in the field to screen for illicit drugs. Some of the most widely used color tests include:

- **Marquis Test:** Used to detect opiates and amphetamines, the Marquis test produces a purple color in the presence of heroin and morphine, and an orange-brown color in the presence of amphetamines.
- **Cobalt Thiocyanate Test:** This test is used to identify cocaine by turning blue when the drug is present.
- **Scott Test:** Another test for cocaine, the Scott test involves a series of color changes (blue to pink) to confirm its presence.
- **Duquenois-Levine Test:** This test is used for the detection of cannabis. A purple color indicates the presence of THC, the psychoactive compound in marijuana.

While color tests are useful for providing quick results, their accuracy can be affected by the presence of other substances, and false positives are possible. Therefore, these tests are often used in conjunction with other methods.

Spot Tests

Spot analysis, spot test analysis, or spot test is a chemical test, a simple and efficient technique where analytic assays are executed in only one, or a few drops, of a chemical solution, preferably in a great piece of filter paper, without using any sophisticated instrumentation.

The development and popularization of the test is credited to Fritz Feigl. A spot test or spot assay can also refer to a test often used in microbiology.

Chemistry

The foundations of Feigl's work on spot analysis were the works of Hugo Schiff (the earliest publication about "spot test" was Shiff's detection of uric acid in 1859) and of Christian Friedrich Schonberg and Friedrich Goppelsröder on capillary analysis.

On the occasion of Feigl's 70th birthday the Chemical Society of Midland sponsored a symposium in 1952, attended by 500 scientists from 24 countries, in which all plenary sessions were related to spot tests.

The test uses the qualitative characteristics of colored compounds to account for performed chemical reactions. This technique has been used to develop new quantification methods using modern technology.

Microbiology

A spot assay or spot test can also refer to a specific test in microbiology. This test is often used to check the growth rate of bacterial or yeast cells on different media or to perform serial dilution tests of micro-organisms. Usually a 96-pinner (often called frogger) is used to perform these spot assay. Another application is high-throughput screening, whichoften uses spot assays to determine the growth of eg. mated cells or to check for protein-protein interactions in a yeast two-hybrid test. This is often done with a robot.

2.3.2 Confirmatory Tests

Confirmatory tests are more sophisticated and provide definitive results regarding the identity of a substance. These tests are conducted in laboratory settings using advanced analytical techniques that offer higher specificity and sensitivity than presumptive tests. Confirmatory tests are essential for providing evidence that can stand up to scrutiny in a court of law.

Chromatography

Chromatography is a powerful technique used to separate and identify the individual components of a mixture. It is widely used in forensic chemistry for the analysis of drugs, explosives, and toxic substances. There are several types of chromatography commonly used in forensic labs:

- **Gas Chromatography (GC)**: GC separates volatile compounds in a mixture by passing them through a column with an inert gas as the mobile phase. This technique is often used in conjunction with mass spectrometry (GC-MS) to identify drugs, poisons, and other chemicals with high precision.
- **Liquid Chromatography (LC)**: Similar to GC, liquid chromatography separates compounds in a liquid phase. High-performance liquid chromatography (HPLC) is frequently used to analyze non-volatile compounds such as pharmaceuticals and biological samples.
- **Thin-Layer Chromatography (TLC)**: TLC involves placing a sample on a stationary phase (usually a silica-coated plate) and allowing a solvent to move through it, separating the components based on their affinities for the stationary and mobile phases. TLC is a cost-effective and simple method for drug screening and identifying inks or dyes in questioned documents.

Spectroscopy

Spectroscopic techniques are used to identify chemical compounds based on the interaction of light with matter. These methods are invaluable in forensic chemistry for their ability to provide detailed information about a substance's molecular structure.

- **Infrared Spectroscopy (IR):** IR spectroscopy measures the absorption of infrared light by a substance, producing a spectrum that reflects the molecular vibrations within the sample. Each compound has a unique infrared spectrum, making this technique useful for identifying unknown substances such as drugs or explosives.
- **Ultraviolet-Visible Spectroscopy (UV-Vis):** UV-Vis spectroscopy measures the absorption of ultraviolet and visible light by a sample. It is often used in toxicology to detect and quantify drugs or poisons in biological samples.
- **Raman Spectroscopy:** This technique is used to analyze materials by measuring the scattering of light. It is especially useful for non-destructive testing of trace evidence such as fibers, paints, and drug residues.

Mass Spectrometry (MS)

Mass spectrometry is a highly sensitive and accurate analytical technique that identifies compounds based on the mass-to-charge ratio of their ions. When combined with chromatography (GC-MS or LC-MS), it provides a powerful method for identifying and quantifying a wide range of substances, including drugs, explosives, and environmental toxins.

In forensic investigations, MS is often used for confirming the identity of unknown substances found at crime scenes, and for detecting trace amounts of drugs or chemicals in biological samples.

Electrophoresis

Electrophoresis is a technique used to separate molecules based on their size and charge by applying an electric field. It is most commonly associated with DNA analysis, where it is used to separate DNA fragments in forensic cases involving biological evidence. Capillary electrophoresis (CE) is often employed to analyze small molecules, such as drugs or metabolites, in forensic toxicology.

2.4 APPLICATIONS OF CHEMICAL TESTS IN FORENSIC INVESTIGATIONS

Chemical tests are used in a variety of forensic contexts to help solve crimes and provide evidence in court. Some of the most common applications include:

2.4.1 Drug Identification

One of the primary uses of chemical tests in forensic chemistry is the identification of illegal drugs. Presumptive tests such as color tests and spot tests are frequently used to screen for drugs in the field, while confirmatory tests like GC-MS provide definitive identification in the lab.

The ability to detect and identify drugs is crucial in cases involving drug trafficking, possession, and overdoses.

2.4.2 Toxicology

Forensic toxicology involves the analysis of biological samples to detect and quantify drugs, alcohol, poisons, and other toxic substances. Chemical tests such as chromatography and spectroscopy are essential for determining the presence of these substances in blood, urine, or tissues. Toxicological analysis is particularly important in cases of poisoning, drug-related deaths, and impaired driving.

2.4.3 Explosive Residue Analysis

In investigations involving bombings or arson, forensic chemists use chemical tests to detect and identify explosive residues. Techniques such as chromatography and mass spectrometry are used to analyze residues from debris or clothing, helping investigators determine the type of explosive used and trace it back to its source.

2.4.4 Environmental and Trace Evidence Analysis

Chemical tests are also used to analyze trace evidence such as fibers, paint, glass, and soil. These materials can provide valuable information about a crime scene or help establish links between suspects and victims. Spectroscopic techniques and microscopy are often employed to examine these materials at a molecular level.

2.5 CHALLENGES AND LIMITATIONS OF CHEMICAL TESTS

While chemical tests are highly effective in forensic investigations, they are not without challenges. Presumptive tests can produce false positives, which can mislead investigators if not followed by confirmatory tests. Contamination of evidence is another concern, as it can compromise the accuracy of test results.

Furthermore, the interpretation of chemical data requires expertise, and errors in analysis or presentation of findings in court can impact the outcome of a case.

Additionally, advancements in forensic technology bring new complexities, as forensic chemists must stay updated on the latest methods and tools to ensure accurate results. Legal standards and the admissibility of scientific evidence in court also place pressure on forensic scientists to provide results that are both scientifically valid and legally defensible.

2.6 CONCLUSION

Chemical tests are integral to forensic chemistry, providing the tools necessary for the identification, analysis, and quantification of substances involved in criminal activities.

From presumptive tests used in the field to sophisticated confirmatory tests performed in the lab, these methods are essential for producing reliable forensic evidence.

While challenges exist, ongoing advancements in chemical testing technologies and methodologies continue to strengthen the role of forensic chemistry in the criminal justice system. Through accurate and thorough chemical testing, forensic scientists help ensure that justice is served by uncovering the truth hidden in physical evidence.

CHAPTER 3
THE MICROSCOPE

3.1 UNDERSTANDING THE MICROSCOPE

The microscope is one of the most important tools in forensic chemistry, providing the ability to analyze materials and substances at a microscopic level. Microscopic analysis is a critical part of the forensic investigation process, as it allows forensic scientists to examine trace evidence such as hair, fibers, paint, glass, and biological materials. Through this detailed observation, forensic chemists can identify, compare, and establish links between different pieces of evidence. Over the years, advances in microscopy have revolutionized forensic science, offering a range of techniques and technologies that allow for more precise and comprehensive analyses.

There are many types of microscopes, and they may be grouped in different ways. One way is to describe the method an instrument uses to interact with a sample and produce images, either by sending a beam of light or electrons through a sample in its optical path, by detecting photon emissions from a sample, or by scanning across and a short distance from the surface of a sample using a probe. The most common microscope (and the first to be invented) is the optical microscope, which uses lenses to refract visible light that passed through a thinly sectioned sample to produce an observable image. Other major types of microscopes are the fluorescence microscope, electron microscope (both the transmission electron microscope and the scanning electron microscope) and various types of scanning probe microscopes.

3.1.1 History

Although objects resembling lenses date back 4,000 years and there are Greek accounts of the optical properties of water-filled spheres (5th century BC) followed by many centuries of writings on optics, the earliest known use of simple microscopes (magnifying glasses) dates back to the widespread use of lenses in eyeglasses in the 13th century. The earliest known examples of compound microscopes, which combine an objective lens near the specimen with an eyepiece to view a real image, appeared in Europe around 1620. The inventor is unknown, even though many claims have been made over the years. Several revolve around the spectacle-making centers in the Netherlands, including claims it was invented in 1590 by Zacharias Janssen (claim made by his son) or Zacharias' father, Hans Martens, or both, claims it was invented by their neighbor and rival spectacle maker, Hans Lippershey (who applied for the first telescope patent in 1608), and claims it was invented by expatriate Cornelis Drebbel, who was noted to have a version in London in 1619. Galileo Galilei (also sometimes cited as compound microscope inventor) seems to have found after 1610 that he could close focus his telescope to view small objects and, after seeing a compound microscope built by Drebbel exhibited in Rome in 1624, built his own improved version. Giovanni Faber coined the name *microscope* for the compound microscope Galileo submitted to the Accademia dei Lincei in 1625 (Galileo had called it the *occhiolino* 'little eye'). René Descartes (*Dioptrique*, 1637) describes microscopes wherein a concave mirror, with its concavity towards the object, is used, in conjunction with a lens, for illuminating the object, which is mounted on a point fixing it at the focus of the mirror.

Rise of Modern Light Microscopes

The first detailed account of the microscopic anatomy of organic tissue based on the use of a microscope did not appear until 1644, in Giambattista Odierna's *L'occhio della mosca*, or *The Fly's Eye*.

The microscope was still largely a novelty until the 1660s and 1670s when naturalists in Italy, the Netherlands and England began using them to study biology. Italian scientist Marcello Malpighi, called the father of histology by some historians of biology, began his analysis of biological structures with the lungs. The publication in 1665 of Robert Hooke's *Micrographia* had a huge impact, largely because of its impressive illustrations. Hooke created tiny lenses of small glass globules made by fusing the ends of threads of spun glass. A significant contribution came from Antonie van Leeuwenhoek who achieved up to 300 times magnification using a simple single lens microscope. He sandwiched a very small glass ball lens between the holes in two metal plates riveted together, and with an adjustable-by-screws needle attached to mount the specimen. Then, Van Leeuwenhoek re-discovered red blood cells (after Jan Swammerdam) and spermatozoa, and helped popularise the use of microscopes to view biological ultrastructure. On 9 October 1676, van Leeuwenhoek reported the discovery of micro-organisms.

The performance of a compound light microscope depends on the quality and correct use of the condensor lens system to focus light on the specimen and the objective lens to capture the light from the specimen and form an image. Early instruments were limited until this principle was fully appreciated and developed from the late 19th to very early 20th century, and until electric lamps were available as light sources. In 1893 August Köhler developed a key principle of sample illumination, Köhler illumination, which is central to achieving the theoretical limits of resolution for the light microscope. This method of sample illumination produces even lighting and overcomes the limited contrast and resolution imposed by early techniques of sample illumination. Further developments in sample illumination came from the discovery of phase contrast by Frits Zernike in 1953, and differential interference contrast illumination by Georges Nomarski in 1955; both of which allow imaging of unstained, transparent samples.

Electron Microscopes

An electron microscope is a microscope that uses a beam of electrons as a source of illumination. They use electron optics that are analogous to the glass lenses of an optical light microscope to control the electron beam, for instance focusing them to produce magnified images or electron diffraction patterns. As the wavelength of an electron can be up to 100,000 times smaller than that of visible light, electron microscopes have a much higher resolution of about 0.1 nm, which compares to about 200 nm for light microscopes. *Electron microscope* may refer to:

- Transmission electron microscopy (TEM) where swift electrons go through a thin sample.
- Scanning transmission electron microscopy (STEM) which is similar to TEM with a scanned electron probe.
- Scanning electron microscope (SEM) which is similar to STEM, but with thick samples.
- Electron microprobe similar to a SEM, but more for chemical analysis.
- Low-energy electron microscopy (LEEM), used to image surfaces.
- Photoemission electron microscopy (PEEM) which is similar to LEEM using electrons emitted from surfaces by photons.

History

Many developments laid the groundwork of the electron optics used in microscopes. One significant step was the work of Hertz in 1883 who made a cathode-ray tube with electrostatic and magnetic deflection, demonstrating manipulation of the direction of an electron beam. Others were focusing of the electrons by an axial magnetic field by Emil Wiechert in 1899, improved oxide-coated cathodes which produced more electrons by Arthur Wehnelt in 1905 and the development of the electromagnetic lens in 1926 by Hans Busch. According to Dennis Gabor, the physicist Leó Szilárd tried in 1928 to convince him to build an electron microscope, for which Szilárd had filed a patent.

To this day the issue of who invented the transmission electron microscope is controversial. In 1928, at the Technische Hochschule in Charlottenburg (now Technische Universität Berlin), Adolf Matthias (Professor of High Voltage Technology and Electrical Installations) appointed Max Knoll to lead a team of researchers to advance research on electron beams and cathode-ray oscilloscopes. The team consisted of several PhD students including Ernst Ruska. In 1931, Max Knoll and Ernst Ruska successfully generated magnified images of mesh grids placed over an anode aperture. The device, a replicate of which is shown in the figure, used two magnetic lenses to achieve higher magnifications, the first electron microscope. (Max Knoll died in 1969, so did not receive a share of the 1986 Nobel prize for the invention of electron microscopes.)

Apparently independent of this effort was work at Siemens-Schuckert by Reinhold Rüdenberg. According to patent law (U.S. Patent No. 2058914 and 2070318, both filed in 1932), he is the inventor of the electron microscope, but it is not clear when he had a working instrument. He stated in a very brief article in 1932 that Siemens had been working on this for some years before the patents were filed in 1932, claiming that his effort was parallel to the university development. He died in 1961, so similar to Max Knoll, was not eligible for a share of the 1986 Nobel prize.

In the following year, 1933, Ruska and Knoll built the first electron microscope that exceeded the resolution of an optical (light) microscope. Four years later, in 1937, Siemens financed the work of Ernst Ruska and Bodo von Borries, and employed Helmut Ruska, Ernst's brother, to develop applications for the microscope, especially with biological specimens. Also in 1937, Manfred von Ardenne pioneered the scanning electron microscope. Siemens produced the first commercial electron microscope in 1938.

The first North American electron microscopes were constructed in the 1930s, at the Washington State University by Anderson and Fitzsimmons and at the University of Toronto by Eli Franklin Burton and students Cecil Hall, James Hillier, and Albert Prebus. Siemens produced a transmission electron microscope (TEM) in 1939. Although current transmission electron microscopes are capable of two million times magnification, as scientific instruments they remain similar but with improved optics.

In the 1940s, high-resolution electron microscopes were developed, enabling greater magnification and resolution. By 1965, Albert Crewe at the University of Chicago introduced the scanning transmission electron microscope using a field emission source, enabling scanning microscopes at high resolution.

By the early 1980s improvements in mechanical stability as well as the use of higher accelerating voltages enabled imaging of materials at the atomic scale. In the 1980s, the field emission gun became common for electron microscopes, improving the image quality due to the additional coherence and lower chromatic aberrations. The 2000s were marked by advancements in aberration-corrected electron microscopy, allowing for significant improvements in resolution and clarity of images.

Transmission Electron Microscope (TEM)

The original form of the electron microscope, the transmission electron microscope (TEM), uses a high voltage electron beam to illuminate the specimen and create an image. An electron beam is produced by an electron gun, with the electrons typically having energies in the range 20 to 400 keV, focused by electromagnetic lenses, and transmitted through the specimen. When it emerges from the specimen, the electron beam carries information about the structure of the specimen that is magnified by lenses of the microscope. The spatial variation in this information (the "image") may be viewed by projecting the magnified electron image onto a detector.

For example, the image may be viewed directly by an operator using a fluorescent viewing screen coated with a phosphor or scintillator material such as zinc sulfide. A high-resolution phosphor may also be coupled by means of a lens optical system or a fibre optic light-guide to the sensor of a digital camera. Direct electron detectors have no scintillator and are directly exposed to the electron beam, which addresses some of the limitations of scintillator-coupled cameras.

The resolution of TEMs is limited primarily by spherical aberration, but a new generation of hardware correctors can reduce spherical aberration to increase the resolution in high-resolution transmission electron microscopy (HRTEM) to below 0.5 angstrom (50 picometres), enabling magnifications above 50 million times. The ability of HRTEM to determine the positions of atoms within materials is useful for nano-technologies research and development.

Scanning Transmission Electron Microscope (STEM)

The STEM rasters a focused incident probe across a specimen. The high resolution of the TEM is thus possible in STEM. The focusing action (and aberrations) occur before the electrons hit the specimen in the STEM, but afterward in the TEM. The STEMs use of SEM-like beam rastering simplifies annular dark-field imaging, and other analytical techniques, but also means that image data is acquired in serial rather than in parallel fashion.

Scanning Electron Microscope (SEM)

The SEM produces images by probing the specimen with a focused electron beam that is scanned across the specimen (raster scanning). When the electron beam interacts with the specimen, it loses energy by a variety of mechanisms. These interactions lead to, among other events, emission of low-energy secondary electrons and high-energy backscattered electrons, light emission (cathodoluminescence) or X-ray emission, all of which provide signals carrying information about the properties of the specimen surface, such as its topography and composition. The image displayed by SEM represents the varying intensity of any of these signals into the image in a position corresponding to the position of the beam on the specimen when the signal was generated.

SEMs are different from TEMs in that they use electrons with much lower energy, generally below 20 keV, while TEMs generally use electrons with energies in the range of 80-300 keV. Thus, the electron sources and optics of the two microscopes have different designs, and they are normally separate instruments.

Electron Diffraction

Transmission electron microscopes can be used in electron diffraction mode where a map of the angles of the electrons leaving the sample is produced. The advantages of electron diffraction over X-ray crystallography are primarily in the size of the crystals. In X-ray crystallography, crystals are commonly visible by the naked eye and are generally in the hundreds of micrometers in length. In comparison, crystals for electron diffraction must be less than a few hundred nanometers in thickness, and have no lower boundary of size. Additionally, electron diffraction is done on a TEM, which can also be used to obtain many other types of information, rather than requiring a separate instrument.

Scanning Electron Microscope (SEM)

To prevent charging and enhance the signal in SEM, non-conductive samples (e.g. biological samples as in figure) can be sputter-coated in a thin film of metal.

Transmission Electron Microscope (TEM)

Materials to be viewed in a transmission electron microscope may require processing to produce a suitable sample. The technique required varies depending on the specimen and the analysis required:

- **Chemical Fixation:** For biological specimens this aims to stabilize the specimen's mobile macromolecular structure by chemical crosslinking of proteins with aldehydes such as formaldehyde and glutaraldehyde, and lipids with osmium tetroxide.
- **Cryofixation:** Freezing a specimen so that the water forms vitreous (non-crystalline) ice. This preserves the specimen in a snapshot of its native state. Methods to achieve this vitrification include plunge freezing rapidly in liquid ethane, and high pressure freezing. An entire field called cryo-electron microscopy has branched from this technique. With the development of cryo-electron microscopy of vitreous sections (CEMOVIS) and cryo-focused ion beam milling of lamellae, it is now possible to observe samples from virtually any biological specimen close to its native state.
- **Dehydration:** Replacement of water with organic solvents such as

ethanol or acetone, followed by critical point drying or infiltration with embedding resins.

- **Embedding, Biological Specimens:** After dehydration, tissue for observation in the transmission electron microscope is embedded so it can be sectioned ready for viewing. To do this the tissue is passed through a 'transition solvent' such as propylene oxide (epoxypropane) or acetone and then infiltrated with an epoxy resin such as Araldite, Epon, or Durcupan; tissues may also be embedded directly in water-miscible acrylic resin. After the resin has been polymerized (hardened) the sample is sectioned by ultramicrotomy and stained.
- **Embedding Materials:** After embedding in resin, the specimen is usually ground and polished to a mirror-like finish using ultra-fine abrasives.
- **Freeze-Fracture or Freeze-Etch:** A preparation method particularly useful for examining lipid membranes and their incorporated proteins in "face on" view. The fresh tissue or cell suspension is frozen rapidly (cryofixation), then fractured by breaking (or by using a microtome) while maintained at liquid nitrogen temperature. The cold fractured surface (sometimes "etched" by increasing the temperature to about -100 °C for several minutes to let some ice sublime) is then shadowed with evaporated platinum or gold at an average angle of 45° in a high vacuum evaporator. The second coat of carbon, evaporated perpendicular to the average surface plane is often performed to improve the stability of the replica coating. The specimen is returned to room temperature and pressure, then the extremely fragile "pre-shadowed" metal replica of the fracture surface is released from the underlying biological material by careful chemical digestion with acids, hypochlorite solution or SDS detergent. The still-floating replica is thoroughly washed free from residual chemicals, carefully fished up on fine grids, dried then viewed in the TEM.
- **Freeze-Fracture Replica Immunogold Labeling (FRIL):** The freeze-fracture method has been modified to allow the identification of the components of the fracture face by immunogold labeling. Instead of removing all the underlying tissue of the thawed replica as the final step before viewing in the microscope the tissue thickness is minimized during or after the fracture process. The thin layer of tissue remains bound to the metal replica so it can be immunogold labeled with antibodies to the structures of choice. The thin layer of the original specimen on the replica with gold attached allows the identification of

structures in the fracture plane. There are also related methods which label the surface of etched cells and other replica labeling variations.

- **Ion Beam Milling:** Thins samples until they are transparent to electrons by firing ions (typically argon) at the surface from an angle and sputtering material from the surface. A subclass of this is focused ion beam milling, where gallium ions are used to produce an electron transparent membrane or 'lamella' in a specific region of the sample, for example through a device within a microprocessor or a focused ion beam SEM. Ion beam milling may also be used for cross-section polishing prior to analysis of materials that are difficult to prepare using mechanical polishing.
- **Negative Stain:** Suspensions containing nanoparticles or fine biological material (such as viruses and bacteria) are briefly mixed with a dilute solution of an electron-opaque solution such as ammonium molybdate, uranyl acetate (or formate), or phosphotungstic acid. This mixture is applied to an EM grid, pre-coated with a plastic film such as formvar, blotted, then allowed to dry. Viewing of this preparation in the TEM should be carried out without delay for best results. The method is important in microbiology for fast but crude morphological identification, but can also be used as the basis for high-resolution 3D reconstruction using EM tomography methodology when carbon films are used for support.
- **Sectioning:** Produces thin slices of the specimen, semitransparent to electrons. These can be cut using ultramicrotomy on an ultramicrotome with a glass or diamond knife to produce ultra-thin sections about 60–90 nm thick. Disposable glass knives are also used because they can be made in the lab and are much cheaper. Sections can also be created in situ by milling in a focused ion beam SEM, where the section is known as a lamella.
- **Staining:** Uses heavy metals such as lead, uranium or tungsten to scatter imaging electrons and thus give contrast between different structures, since many (especially biological) materials are nearly "transparent" to electrons (weak phase objects). In biology, specimens can be stained "en bloc" before embedding and also later after sectioning. Typically thin sections are stained for several minutes with an aqueous or alcoholic solution of uranyl acetate followed by aqueous lead citrate.

EM Workflows

In their most common configurations, electron microscopes produce images with a single brightness value per pixel, with the results usually rendered in greyscale.

However, often these images are then colourized through the use of feature-detection software, or simply by hand-editing using a graphics editor. This may be done to clarify structure or for aesthetic effect and generally does not add new information about the specimen.

Electron microscopes are now frequently used in more complex workflows, with each workflow typically using multiple technologies to enable more complex and/or more quantitative analyses of a sample. A few examples are outlined below, but this should not be considered an exhaustive list. The choice of workflow will be highly dependent on the application and the requirements of the corresponding scientific questions, such as resolution, volume, nature of the target molecule, etc.

For example, images from light and electron microscopy of the same region of a sample can be overlaid to correlate the data from the two modalities. This is commonly used to provide higher resolution contextual EM information about a fluorescently labelled structure. This correlative light and electron microscopy (CLEM) is one of a range of correlative workflows now available. Another example is high resolution mass spectrometry (ion microscopy), which has been used to provide correlative information about subcellular antibiotic localisation, data that would be difficult to obtain by other means.

The initial role of electron microscopes in imaging two-dimensional slices (TEM) or a specimen surface (SEM with secondary electrons) has also increasingly expanded into the depth of samples. An early example of these 'volume EM' workflows was simply to stack TEM images of serial sections cut through a sample. The next development was virtual reconstruction of a thick section (200-500 nm) volume by backprojection of a set of images taken at different tilt angles - TEM tomography.

Serial Imaging for Volume EM

To acquire volume EM datasets of larger depths than TEM tomography (micrometers or millimeters in the z axis), a series of images taken through the sample depth can be used. For example, ribbons of serial sections can be imaged in a TEM as described above, and when thicker sections are used, serial TEM tomography can be used to increase the z-resolution. More recently, back scattered electron (BSE) images can be acquired of a larger series of sections collected on silicon wafers, known as SEM array tomography. An alternative approach is to use BSE SEM to image the block surface instead of the section, after each section has been removed. By this method, an ultramicrotome installed in an SEM chamber can increase automation of the workflow; the specimen block is loaded in the chamber and the system programmed to continuously cut and image through the sample. This is known as serial block face SEM. A related method uses focused ion beam milling instead of an ultramicrotome to remove sections. In these serial imaging methods, the output is essentially a sequence of images through a specimen block that can be digitally aligned in sequence and thus reconstructed into a volume EM dataset. The increased volume available in these methods has expanded the capability of electron microscopy to address new questions, such as mapping neural connectivity in the brain, and membrane contact sites between organelles.

Disadvantages

Electron microscopes are expensive to build and maintain. Microscopes designed to achieve high resolutions must be housed in stable buildings (sometimes underground) with special services such as magnetic field canceling systems.

The samples largely have to be viewed in vacuum, as the molecules that make up air would scatter the electrons. An exception is liquid-phase electron microscopy using either a closed liquid cell or an environmental chamber, for example, in the environmental scanning electron microscope, which allows hydrated samples to be viewed in a low-pressure (up to 20 Torr or 2.7 kPa) wet environment. Various techniques for in situ electron microscopy of gaseous samples have been developed.

Scanning electron microscopes operating in conventional high-vacuum mode usually image conductive specimens; therefore non-conductive materials require conductive coating (gold/palladium alloy, carbon, osmium, etc.). The low-voltage mode of modern microscopes makes possible the observation of non-conductive specimens without coating. Non-conductive materials can be imaged also by a variable pressure (or environmental) scanning electron microscope.

Small, stable specimens such as carbon nanotubes, diatom frustules and small mineral crystals (asbestos fibres, for example) require no special treatment before being examined in the electron microscope. Samples of hydrated materials, including almost all biological specimens, have to be prepared in various ways to stabilize them, reduce their thickness (ultrathin sectioning) and increase their electron optical contrast (staining). These processes may result in *artifacts*, but these can usually be identified by comparing the results obtained by using radically different specimen preparation methods. Since the 1980s, analysis of cryofixed, vitrified specimens has also become increasingly used by scientists, further confirming the validity of this technique.

Scanning Probe Microscopes

Scanning probe microscopy (SPM) is a branch of microscopy that forms images of surfaces using a physical probe that scans the specimen. SPM was founded in 1981, with the invention of the scanning tunneling microscope, an instrument for imaging surfaces at the atomic level. The first successful scanning tunneling microscope experiment was done by Gerd Binnig and Heinrich Rohrer. The key to their success was using a feedback loop to regulate gap distance between the sample and the probe.

Many scanning probe microscopes can image several interactions simultaneously. The manner of using these interactions to obtain an image is generally called a mode. The resolution varies somewhat from technique to technique, but some probe techniques reach a rather impressive atomic resolution. This is due largely because piezoelectric actuators can execute motions with a precision and accuracy at the atomic level or better on electronic command. This family of techniques can be called "piezoelectric techniques". The other common denominator is that the data are typically obtained as a two-dimensional grid of data points, visualized in false color as a computer image.

Image Formation

To form images, scanning probe microscopes raster scan the tip over the surface. At discrete points in the raster scan a value is recorded (which value depends on the type of SPM and the mode of operation). These recorded values are displayed as a heat map to produce the final STM images, usually using a black and white or an orange color scale.

Constant Interaction Mode

In constant interaction mode (often referred to as "in feedback"), a feedback loop is used to physically move the probe closer to or further from the surface (in the *z* axis) under study to maintain a constant interaction.

This interaction depends on the type of SPM, for scanning tunneling microscopy the interaction is the tunnel current, for contact mode AFM or MFM it is the cantilever deflection, etc. The type of feedback loop used is usually a PI-loop, which is a PID-loop where the differential gain has been set to zero (as it amplifies noise). The *z* position of the tip (scanning plane is the *xy*-plane) is recorded periodically and displayed as a heat map. This is normally referred to as a topography image.

In this mode a second image, known as the "error signal" or "error image" is also taken, which is a heat map of the interaction which was fed back on. Under perfect operation this image would be a blank at a constant value which was set on the feedback loop. Under real operation the image shows noise and often some indication of the surface structure. The user can use this image to edit the feedback gains to minimise features in the error signal.

If the gains are set incorrectly, many imaging artifacts are possible. If gains are too low features can appear smeared. If the gains are too high the feedback can become unstable and oscillate, producing striped features in the images which are not physical.

Constant Height Mode

In constant height mode the probe is not moved in the *z*-axis during the raster scan. Instead the value of the interaction under study is recorded (i.e. the tunnel current for STM, or the cantilever oscillation amplitude for amplitude modulated non-contact AFM). This recorded information is displayed as a heat map, and is usually referred to as a constant height image.

Constant height imaging is much more difficult than constant interaction imaging as the probe is much more likely to crash into the sample surface. Usually before performing constant height imaging one must image in constant interaction mode to check the surface has no large contaminants in the imaging region, to measure and correct for the sample tilt, and (especially for slow scans) to measure and correct for thermal drift of the sample. Piezoelectric creep can also be a problem, so the microscope often needs time to settle after large movements before constant height imaging can be performed. Constant height imaging can be advantageous for eliminating the possibility of feedback artifacts.

Probe Tips

The nature of an SPM probe tip depends entirely on the type of SPM being used. The combination of tip shape and topography of the sample make up a SPM image. However, certain characteristics are common to all, or at least most, SPMs.

Most importantly the probe must have a very sharp apex. The apex of the probe defines the resolution of the microscope, the sharper the probe the better the resolution. For atomic resolution imaging the probe must be terminated by a single atom.

For many cantilever based SPMs (e.g. AFM and MFM), the entire cantilever and integrated probe are fabricated by acid [etching], usually from silicon nitride. Conducting probes, needed for STM and SCM among others, are usually constructed from platinum/iridium wire for ambient operations, or tungsten for UHV operation. Other materials such as gold are sometimes used either for sample specific reasons or if the SPM is to be combined with other experiments such as TERS. Platinum/iridium (and other ambient) probes are normally cut using sharp wire cutters, the optimal method is to cut most of the way through the wire and then pull to snap the last of the wire, increasing the likelihood of a single atom termination. Tungsten wires are usually electrochemically etched, following this the oxide layer normally needs to be removed once the tip is in UHV conditions.

It is not uncommon for SPM probes (both purchased and "home-made") to not image with the desired resolution. This could be a tip which is too blunt or the probe may have more than one peak, resulting in a doubled or ghost image. For some probes, *in situ* modification of the tip apex is possible, this is usually done by either crashing the tip into the surface or by applying a large electric field. The latter is achieved by applying a bias voltage (of order 10V) between the tip and the sample, as this distance is usually 1-3 Angstroms, a very large field is generated.

The additional attachment of a quantum dot to the tip apex of a conductive probe enables surface potential imaging with high lateral resolution, scanning quantum dot microscopy.

Advantages

- The resolution of the microscopes is not limited by diffraction, only by the size of the probe-sample interaction volume (i.e., point spread function), which can be as small as a few picometres. Hence the ability to measure small local differences in object height (like that of 135 picometre steps on <100> silicon) is unparalleled. Laterally the probe-sample interaction extends only across the tip atom or atoms involved in the interaction.
- The interaction can be used to modify the sample to create small structures (Scanning probe lithography).
- Unlike electron microscope methods, specimens do not require a partial vacuum but can be observed in air at standard temperature and pressure or while submerged in a liquid reaction vessel.

Disadvantages

- The detailed shape of the scanning tip is sometimes difficult to determine. Its effect on the resulting data is particularly noticeable if the specimen varies greatly in height over lateral distances of 10 nm or less.
- The scanning techniques are generally slower in acquiring images, due to the scanning process. As a result, efforts are being made to greatly improve the scanning rate. Like all scanning techniques, the embedding of spatial information into a time sequence opens the door to uncertainties in metrology, say of lateral spacings and angles, which arise due to time-domain effects like specimen drift, feedback loop oscillation, and mechanical vibration.
- The maximum image size is generally smaller.
- Scanning probe microscopy is often not useful for examining buried solid-solid or liquid-liquid interfaces.

Scanning Photo Current Microscopy (SPCM)

SPCM can be considered as a member of the Scanning Probe Microscopy (SPM) family. The difference between other SPM techniques and SPCM is, it exploits a focused laser beam as the local excitation source instead of a probe tip.

Characterization and analysis of spatially resolved optical behavior of materials is very important in opto-electronic industry. Simply this involves studying how the properties of a material vary across its surface or bulk structure. Techniques that enable spatially resolved optoelectronic measurements provide valuable insights for the enhancement of optical performance. Scanning electron microscopy (SPCM) has emerged as a powerful technique which can investigate spatially resolved optoelectronic properties in semiconductor nano structures.

Principle

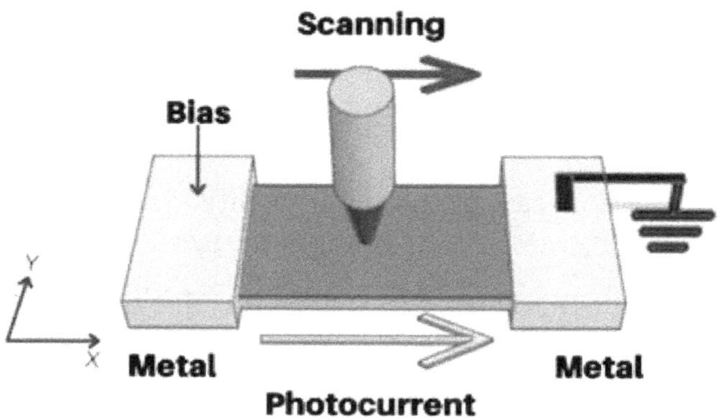

Fig 1: Laser Scan of the Scanning Photocurrent Microscope

In SPCM, a focused laser beam is used to excite the semiconducting material producing excitons (electro-hole pairs). These excitons undergo different mechanisms and if they can reach the nearby electrodes before the recombination takes place a photocurrent is generated. This photocurrent is position dependent as it, raster scans the device.

SPCM Analysis

Using the position dependent photocurrent map, important photocurrent dynamics can be analyzed. SPCM provides information such as characteristic length such as minority diffusion length, recombination dynamics, doping concentration, internal electric field, etc.

Visualization and Analysis Software

In all instances and contrary to optical microscopes, rendering software is necessary to produce images. Such software is produced and embedded by instrument manufacturers but also available as an accessory from specialized work groups or companies. The main packages used are freeware: Gwyddion, WSxM (developed by Nanotec) and commercial: SPIP (developed by Image Metrology), FemtoScan Online (developed by Advanced Technologies Center), MountainsMap SPM (developed by Digital Surf), TopoStitch (developed by Image Metrology).

Fluorescence Microscopes

A fluorescence microscope is an optical microscope that uses fluorescence instead of, or in addition to, scattering, reflection, and attenuation or absorption, to study the properties of organic or inorganic substances. "Fluorescence microscope" refers to any microscope that uses fluorescence to generate an image, whether it is a simple set up like an epifluorescence microscope or a more complicated design such as a confocal microscope, which uses optical sectioning to get better resolution of the fluorescence image.

Principle

The specimen is illuminated with light of a specific wavelength (or wavelengths) which is absorbed by the fluorophores, causing them to emit light of longer wavelengths (i.e., of a different color than the absorbed light). The illumination light is separated from the much weaker emitted fluorescence through the use of a spectral emission filter. Typical components of a fluorescence microscope are a light source (xenon arc lamp or mercury-vapor lamp are common; more advanced forms are high-power LEDs and lasers), the excitation filter, the dichroic mirror (or dichroic beamsplitter), and the emission filter. The filters and the dichroic beamsplitter are chosen to match the spectral excitation and emission characteristics of the fluorophore used to label the specimen. In this manner, the distribution of a single fluorophore (color) is imaged at a time. Multi-color images of several types of fluorophores must be composed by combining several single-color images.

Most fluorescence microscopes in use are epifluorescence microscopes, where excitation of the fluorophore and detection of the fluorescence are done through the same light path (i.e. through the objective). These microscopes are widely used in biology and are the basis for more advanced microscope designs, such as the confocal microscope and the total internal reflection fluorescence microscope (TIRF).

Epifluorescence Microscopy

The majority of fluorescence microscopes, especially those used in the life sciences, are of the epifluorescence design shown in the diagram. Light of the excitation wavelength illuminates the specimen through the objective lens. The fluorescence emitted by the specimen is focused to the detector by the same objective that is used for the excitation which for greater resolution will need objective lens with higher numerical aperture. Since most of the excitation light is transmitted through the specimen, only reflected excitatory light reaches the objective together with the emitted light and the epifluorescence method therefore gives a high signal-to-noise ratio. The dichroic beamsplitter acts as a wavelength specific filter, transmitting fluoresced light through to the eyepiece or detector, but reflecting any remaining excitation light back towards the source.

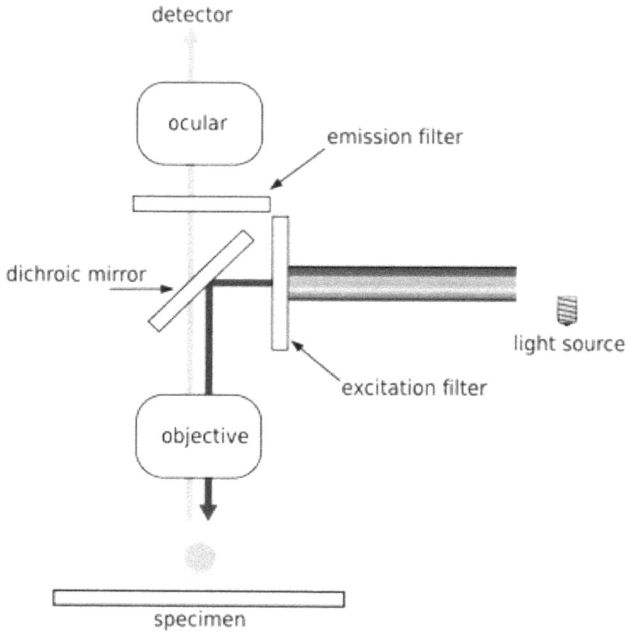

Fig 2: Schematic of a Fluorescence Microscope

Light Sources

Fluorescence microscopy requires intense, near-monochromatic, illumination which some widespread light sources, like halogen lamps cannot provide. Four main types of light source are used, including xenon arc lamps or mercury-vapor lamps with an excitation filter, lasers, supercontinuum sources, and high-power LEDs. Lasers are most widely used for more complex fluorescence microscopy techniques like confocal microscopy and total internal reflection fluorescence microscopy while xenon lamps, and mercury lamps, and LEDs with a dichroic excitation filter are commonly used for widefield epifluorescence microscopes. By placing two microlens arrays into the illumination path of a widefield epifluorescence microscope, highly uniform illumination with a coefficient of variation of 1-2% can be achieved.

Sample Preparation

In order for a sample to be suitable for fluorescence microscopy it must be fluorescent. There are several methods of creating a fluorescent sample; the main techniques are labelling with fluorescent stains or, in the case of biological samples, expression of a fluorescent protein. Alternatively the intrinsic fluorescence of a sample (i.e., autofluorescence) can be used. In the life sciences fluorescence microscopy is a powerful tool which allows the specific and sensitive staining of a specimen in order to detect the distribution of proteins or other molecules of interest. As a result, there is a diverse range of techniques for fluorescent staining of biological samples.

Biological Fluorescent Stains

Many fluorescent stains have been designed for a range of biological molecules. Some of these are small molecules which are intrinsically fluorescent and bind a biological molecule of interest. Major examples of these are nucleic acid stains such as DAPI and Hoechst (excited by UV wavelength light) and DRAQ5 and DRAQ7 (optimally excited by red light) which all bind the minor groove of DNA, thus labeling the nuclei of cells. Others are drugs, toxins, or peptides which bind specific cellular structures and have been derivatised with a fluorescent reporter. A major example of this class of fluorescent stain is phalloidin, which is used to stain actin fibers in mammalian cells. A new peptide, known as the Collagen Hybridizing Peptide, can also be conjugated with fluorophores and used to stain *denatured* collagen fibers. Staining of the plant cell walls is performed using stains or dyes that bind cellulose or pectin. The quest for fluorescent probes with a high specificity that also allow live imaging of plant cells is ongoing. There are many fluorescent molecules called fluorophores or fluorochromes such as fluorescein, Alexa Fluors, or DyLight 488, which can be chemically linked to a different molecule which binds the target of interest within the sample.

Immunofluorescence

Immunofluorescence is a technique which uses the highly specific binding of an antibody to its antigen in order to label specific proteins or other molecules within the cell. A sample is treated with a primary antibody specific for the molecule of interest. A fluorophore can be directly conjugated to the primary antibody. Alternatively a secondary antibody, conjugated to a fluorophore, which binds specifically to the first antibody can be used. For example, a primary antibody raised in a mouse which recognises tubulin combined with a secondary anti-mouse antibody derivatised with a fluorophore could be used to label microtubules in a cell.

Fluorescent Proteins

The modern understanding of genetics and the techniques available for modifying DNA allow scientists to genetically modify proteins to also carry a fluorescent protein reporter. In biological samples this allows a scientist to directly make a protein of interest fluorescent. The protein location can then be directly tracked, including in live cells.

Limitations

Fluorophores lose their ability to fluoresce as they are illuminated in a process called photobleaching. Photobleaching occurs as the fluorescent molecules accumulate chemical damage from the electrons excited during fluorescence. Photobleaching can severely limit the time over which a sample can be observed by fluorescence microscopy. Several techniques exist to reduce photobleaching such as the use of more robust fluorophores, by minimizing illumination, or by using photoprotective scavenger chemicals.

Fluorescence microscopy with fluorescent reporter proteins has enabled analysis of live cells by fluorescence microscopy, however cells are susceptible to phototoxicity, particularly with short wavelength light. Furthermore, fluorescent molecules have a tendency to generate reactive chemical species when under illumination which enhances the phototoxic effect.

Unlike transmitted and reflected light microscopy techniques, fluorescence microscopy only allows observation of the specific structures which have been labeled for fluorescence. For example, observing a tissue sample prepared with a fluorescent DNA stain by fluorescence microscopy only reveals the organization of the DNA within the cells and reveals nothing else about the cell morphologies.

Computational techniques that propose to estimate the fluorescent signal from non-fluorescent images (such as brightfield) may reduce these concerns. In general, these approaches involve training a deep convolutional neural network on stained cells and then estimating the fluorescence on unstained samples. Thus by decoupling the cells under investigation from the cells used to train the network, imaging can performed quicker and with reduced phototoxicity.

Sub-Diffraction Techniques

The wave nature of light limits the size of the spot to which light can be focused due to the diffraction limit. This limitation was described in the 19th century by Ernst Abbe and "limits an optical microscope's resolution to approximately half of the wavelength of the light used." Fluorescence microscopy is central to many techniques which aim to reach past this limit by specialized optical configurations.

Several improvements in microscopy techniques have been invented in the 20th century and have resulted in increased resolution and contrast to some extent. However they did not overcome the diffraction limit. In 1978 first theoretical ideas have been developed to break this barrier by using a 4Pi microscope as a confocal laser scanning fluorescence microscope where the light is focused ideally from all sides to a common focus which is used to scan the object by 'point-by-point' excitation combined with 'point-by-point' detection. However, the first experimental demonstration of the 4pi microscope took place in 1994. 4Pi microscopy maximizes the amount of available focusing directions by using two opposing objective lenses or two-photon excitation microscopy using redshifted light and multi-photon excitation.

Integrated correlative microscopy combines a fluorescence microscope with an electron microscope. This allows one to visualize ultrastructure and contextual information with the electron microscope while using the data from the fluorescence microscope as a labelling tool.

The first technique to really achieve a sub-diffraction resolution was STED microscopy, proposed in 1994. This method and all techniques following the RESOLFT concept rely on a strong non-linear interaction between light and fluorescing molecules. The molecules are driven strongly between distinguishable molecular states at each specific location, so that finally light can be emitted at only a small fraction of space, hence an increased resolution.

As well in the 1990s another super resolution microscopy method based on wide field microscopy has been developed. Substantially improved size resolution of cellular nanostructures stained with a fluorescent marker was achieved by development of SPDM localization microscopy and the structured laser illumination (spatially modulated illumination, SMI). Combining the principle of SPDM with SMI resulted in the development of the Vertico SMI microscope. Single molecule detection of normal blinking fluorescent dyes like green fluorescent protein (GFP) can be achieved by using a further development of SPDM the so-called SPDMphymod technology which makes it possible to detect and count two different fluorescent molecule types at the molecular level (this technology is referred to as two-color localization microscopy or 2CLM).

Alternatively, the advent of photoactivated localization microscopy could achieve similar results by relying on blinking or switching of single molecules, where the fraction of fluorescing molecules is very small at each time. This stochastic response of molecules on the applied light corresponds also to a highly nonlinear interaction, leading to subdiffraction resolution.

Super Resolution Microscopes

Much current research (in the early 21st century) on optical microscope techniques is focused on development of superresolution analysis of fluorescently labelled samples. Structured illumination can improve resolution by around two to four times and techniques like stimulated emission depletion (STED) microscopy are approaching the resolution of electron microscopes. This occurs because the diffraction limit is occurred from light or excitation, which makes the resolution must be doubled to become super saturated. Stefan Hell was awarded the 2014 Nobel Prize in Chemistry for the development of the STED technique, along with

Eric Betzig and William Moerner who adapted fluorescence microscopy for single-molecule visualization.

X-Ray Microscopes

An X-ray microscope uses electromagnetic radiation in the X-ray band to produce magnified images of objects. Since X-rays penetrate most objects, there is no need to specially prepare them for X-ray microscopy observations.

Unlike visible light, X-rays do not reflect or refract easily and are invisible to the human eye. Therefore, an X-ray microscope exposes film or uses a charge-coupled device (CCD) detector to detect X-rays that pass through the specimen. It is a contrast imaging technology using the difference in absorption of soft X-rays in the water window region (wavelengths: 2.34–4.4 nm, energies: 280–530 eV) by the carbon atom (main element composing the living cell) and the oxygen atom (an element of water).

Microfocus X-ray also achieves high magnification by projection. A microfocus X-ray tube produces X-rays from an extremely small focal spot (5 µm down to 0.1 µm). The X-rays are in the more conventional X-ray range (20 to 300 keV) and are not re-focused.

Invention and Development

The history of X-ray microscopy can be traced back to the early 20th century. After the German physicist Röntgen discovered X-rays in 1895, scientists soon illuminated an object using an X-ray point source and captured the shadow images of the object with a resolution of several micrometers. In 1918, Einstein pointed out that the refractive index for X-rays in most mediums should be just slightly greater than 1, which means that refractive optical parts would be difficult to use for X-ray applications.

Early X-ray microscopes by Paul Kirkpatrick and Albert Baez used grazing-incidence reflective X-ray optics to focus the X-rays, which grazed X-rays off parabolic curved mirrors at a very high angle of incidence. An alternative method of focusing X-rays is to use a tiny Fresnel zone plate of concentric gold or nickel rings on a silicon dioxide substrate. Sir Lawrence Bragg produced some of the first usable X-ray images with his apparatus in the late 1940s.

In the 1950s Sterling Newberry produced a shadow X-ray microscope, which placed the specimen between the source and a target plate, this became the basis for the first commercial X-ray microscopes from the General Electric Company.

After a silent period in the 1960s, X-ray microscopy regained people's attention in the 1970s. In 1972, Horowitz and Howell built the first synchrotron-based X-ray microscope at the Cambridge Electron Accelerator. This microscope scanned samples using synchrotron radiation from a tiny pinhole and showed the abilities of both transmission and fluorescence microscopy. Other developments in this period include the first holographic demonstration by Sadao Aoki and Seishi Kikuta in Japan, the first TXMs using zone plates by Schmahl et al., and Stony Brook's experiments in STXM.

The uses of synchrotron light sources brought new possibilities for X-ray microscopy in the 1980s. However, as new synchrotron-source-based microscopes were built in many groups, people realized that it was difficult to perform such experiments due to insufficient technological capabilities at that time, such as poor coherent illuminations, poor-quality x-ray optical elements, and user-unfriendly light sources.

Entering the 1990s, new instruments and new light sources greatly fueled the improvementofX-raymicroscopy.Microscopymethodsincludingtomography,cryo-, and cryo-tomography were successfully demonstrated. With rapid development, X-ray microscopy found new applications in soil science, geochemistry, polymer sciences, and magnetism. The hardware was also miniaturized, so that researchers could perform experiments in their own laboratories.

Extremely high-intensity sources of 9.25 keV X-rays for X-ray phase-contrast microscopy, from a focal spot about 10 μm × 10 μm, may be obtained with a non-synchrotron X-ray source that uses a focused electron beam and a liquid-metal anode. This was demonstrated in 2003 and in 2017 was used to image mouse brain at a voxel size of about one cubic micrometer.

With the applications continuing to grow, X-ray microscopy has become a routine, proven technique used in environmental and soil sciences, geo- and cosmo-chemistry, polymer sciences, biology, magnetism, material sciences. With this increasing demand for X-ray microscopy in these fields, microscopes based on synchrotron, liquid-metal anode, and other laboratory light sources are being built around the world. X-ray optics and components are also being commercialized rapidly.

Advanced Light Source

The Advanced Light Source (ALS) in Berkeley, California, is home to XM-1, a full-field soft X-ray microscope operated by the Center for X-ray Optics and dedicated to various applications in modern nanoscience, such as nanomagnetic materials, environmental and materials sciences and biology. XM-1 uses an X-ray lens to focus X-rays on a CCD, in a manner similar to an optical microscope. XM-1 held the world record in spatial resolution with Fresnel zone plates down to 15 nm and is able to combine high spatial resolution with a sub-100ps time resolution to study e.g. ultrafast spin dynamics. In July 2012, a group at DESY claimed a record spatial resolution of 10 nm, by using the hard X-ray scanning microscope at PETRA III. The ALS is also home to the world's first soft x-ray microscope designed for biological and biomedical research. This new instrument, XM-2 was designed and built by scientists from the National Center for X-ray Tomography. XM-2 is capable of producing 3-dimensional tomograms of cells.

Liquid-Metal-Anode X-Ray Source

Extremely high-intensity sources of 9.25 keV X-rays (gallium K-alpha line) for X-ray phase-contrast microscopy, from a focal spot about 10 um x 10 um, may be obtained with an X-ray source which uses a liquid metal galinstan anode. This was demonstrated in 2003. The metal flows from a nozzle downward at a high speed and the high intensity electron source is focused upon it. The rapid flow of metal carries current, but the physical flow prevents a great deal of anode heating (due to forced-convective heat removal), and the high boiling point of galinstan inhibits vaporization of the anode. The technique has been used to image mouse brain in three dimensions at a voxel size of about one cubic micrometer.

Scanning Transmission

Sources of soft X-rays suitable for microscopy, such as synchrotron radiation sources, have fairly low brightness of the required wavelengths, so an alternative method of image formation is scanning transmission soft X-ray microscopy. Here the X-rays are focused to a point, and the sample is mechanically scanned through the produced focal spot. At each point the transmitted X-rays are recorded using a detector such as a proportional counter or an avalanche photodiode. This type of *scanning transmission X-ray microscope* (STXM) was first developed by researchers at Stony Brook University and was employed at the National Synchrotron Light Source at Brookhaven National Laboratory.

Resolution

The resolution of X-ray microscopy lies between that of the optical microscope and the electron microscope. It has an advantage over conventional electron microscopy is that it can view biological samples in their natural state. Electron microscopy is widely used to obtain images with nanometer to sub-Angstrom level resolution but the relatively thick living cell cannot be observed as the sample has to be chemically fixed, dehydrated, embedded in resin, then sliced ultra thin. However, it should be mentioned that cryo-electron microscopy allows the observation of biological specimens in their hydrated natural state, albeit embedded in water ice. Until now, resolutions of 30 nanometer are possible using the Fresnel zone plate lens which forms the image using the soft x-rays emitted from a synchrotron. Recently, the use of soft x-rays emitted from laser-produced plasmas rather than synchrotron radiation is becoming more popular.

Analysis

Additionally, X-rays cause fluorescence in most materials, and these emissions can be analyzed to determine the chemical elements of an imaged object. Another use is to generate diffraction patterns, a process used in X-ray crystallography. By analyzing the internal reflections of a diffraction pattern (usually with a computer program), the three-dimensional structure of a crystal can be determined down to the placement of individual atoms within its molecules. X-ray microscopes are sometimes used for these analyses because the samples are too small to be analyzed in any other way.

Biological Applications

One early applications of X-ray microscopy in biology was contact imaging, pioneered by Goby in 1913. In this technique, soft x-rays irradiate a specimen and expose the x-ray sensitive emulsions beneath it. Then, magnified tomographic images of the emulsions, which correspond to the x-ray opacity maps of the specimen, are recorded using a light microscope or an electron microscope. A unique advantage that X-ray contact imaging offered over electron microscopy was the ability to image wet biological materials.

Thus, it was used to study the micro and nanoscale structures of plants, insects, and human cells. However, several factors, including emulsion distortions, poor illumination conditions, and low resolutions of ways to examine the emulsions, limit the resolution of contacting imaging. Electron damage of the emulsions and diffraction effects can also result in artifacts in the final images.

X-ray microscopy has its unique advantages in terms of nanoscale resolution and high penetration ability, both of which are needed in biological studies. With the recent significant progress in instruments and focusing, the three classic forms of optics—diffractive, reflective, refractive optics—have all successfully expanded into the X-ray range and have been used to investigate the structures and dynamics at cellular and sub-cellular scales.

In 2005, Shapiro et al. reported cellular imaging of yeasts at a 30 nm resolution using coherent soft X-ray diffraction microscopy. In 2008, X-ray imaging of an unstained virus was demonstrated. A year later, X-ray diffraction was further applied to visualize the three-dimensional structure of an unstained human chromosome. X-ray microscopy has thus shown its great ability to circumvent the diffractive limit of classic light microscopes; however, further enhancement of the resolution is limited by detector pixels, optical instruments, and source sizes.

A longstanding major concern of X-ray microscopy is radiation damage, as high energy X-rays produce strong radicals and trigger harmful reactions in wet specimens. As a result, biological samples are usually fixated or freeze-dried before being irradiated with high-power X-rays. Rapid cryo-treatments are also commonly used in order to preserve intact hydrated structures.

3.1.2 Types

Microscopes can be separated into several different classes. One grouping is based on what interacts with the sample to generate the image, i.e., light or photons (optical microscopes), electrons (electron microscopes) or a probe (scanning probe microscopes). Alternatively, microscopes can be classified based on whether they analyze the sample via a scanning point (confocal optical microscopes, scanning electron microscopes and scanning probe microscopes) or analyze the sample all at once (wide field optical microscopes and transmission electron microscopes).

Wide field optical microscopes and transmission electron microscopes both use the theory of lenses (optics for light microscopes and electromagnet lenses for electron microscopes) in order to magnify the image generated by the passage of a wave transmitted through the sample, or reflected by the sample. The waves used are electromagnetic (in optical microscopes) or electron beams (in electron microscopes). Resolution in these microscopes is limited by the wavelength of the radiation used to image the sample, where shorter wavelengths allow for a higher resolution.

Scanning optical and electron microscopes, like the confocal microscope and scanning electron microscope, use lenses to focus a spot of light or electrons onto the sample then analyze the signals generated by the beam interacting with the sample. The point is then scanned over the sample to analyze a rectangular region. Magnification of the image is achieved by displaying the data from scanning a physically small sample area on a relatively large screen. These microscopes have the same resolution limit as wide field optical, probe, and electron microscopes.

Scanning probe microscopes also analyze a single point in the sample and then scan the probe over a rectangular sample region to build up an image. As these microscopes do not use electromagnetic or electron radiation for imaging they are not subject to the same resolution limit as the optical and electron microscopes described above.

Optical Microscope

The most common type of microscope (and the first invented) is the optical microscope. This is an optical instrument containing one or more lenses producing an enlarged image of a sample placed in the focal plane. Optical microscopes have refractive glass (occasionally plastic or quartz), to focus light on the eye or on to another light detector. Mirror-based optical microscopes operate in the same manner. Typical magnification of a light microscope, assuming visible range light, is up to 1,250× with a theoretical resolution limit of around

0.250 micrometres or 250 nanometres. This limits practical magnification to ~1,500×. Specialized techniques (e.g., scanning confocal microscopy, Vertico SMI) may exceed this magnification but the resolution is diffraction limited. The use of shorter wavelengths of light, such as ultraviolet, is one way to improve the spatial resolution of the optical microscope, as are devices such as the near-field scanning optical microscope.

Sarfus is a recent optical technique that increases the sensitivity of a standard optical microscope to a point where it is possible to directly visualize nanometric films (down to 0.3 nanometre) and isolated nano-objects (down to 2 nm-diameter). The technique is based on the use of non-reflecting substrates for cross-polarized reflected light microscopy.

Ultraviolet light enables the resolution of microscopic features as well as the imaging of samples that are transparent to the eye. Near infrared light can be used to visualize circuitry embedded in bonded silicon devices, since silicon is transparent in this region of wavelengths.

In fluorescence microscopy many wavelengths of light ranging from the ultraviolet to the visible can be used to cause samples to fluoresce, which allows viewing by eye or with specifically sensitive cameras.

Phase-contrast microscopy is an optical microscopic illumination technique in which small phase shifts in the light passing through a transparent specimen are converted into amplitude or contrast changes in the image. The use of phase contrast does not require staining to view the slide. This microscope technique made it possible to study the cell cycle in live cells.

The traditional optical microscope has more recently evolved into the digital microscope. In addition to, or instead of, directly viewing the object through the eyepieces, a type of sensor similar to those used in a digital camera is used to obtain an image, which is then displayed on a computer monitor. These sensors may use CMOS or charge-coupled device (CCD) technology, depending on the application.

Digital microscopy with very low light levels to avoid damage to vulnerable biological samples is available using sensitive photon-counting digital cameras. It has been demonstrated that a light source providing pairs of entangled photons may minimize the risk of damage to the most light-sensitive samples. In this application of ghost imaging to photon-sparse microscopy, the sample is illuminated with infrared photons, each of which is spatially correlated with an entangled partner in the visible band for efficient imaging by a photon-counting camera.

Electron Microscope

The two major types of electron microscopes are transmission electron microscopes (TEMs) and scanning electron microscopes (SEMs). They both have series of electromagnetic and electrostatic lenses to focus a high energy beam of electrons on a sample. In a TEM the electrons pass through the sample, analogous to basic optical microscopy. This requires careful sample preparation, since electrons are scattered strongly by most materials. The samples must also be very thin (below 100 nm) in order for the electrons to pass through it. Cross-sections of cells stained with osmium and heavy metals reveal clear organelle membranes and proteins such as ribosomes. With a 0.1 nm level of resolution, detailed views of viruses (20 – 300 nm) and a strand of DNA (2 nm in width) can be obtained. In contrast, the SEM has raster coils to scan the surface of bulk objects with a fine electron beam. Therefore, the specimen do not necessarily need to be sectioned, but coating with a nanometric metal or carbon layer may be needed for nonconductive samples. SEM allows fast surface imaging of samples, possibly in thin water vapor to prevent drying.

Scanning Probe

The different types of scanning probe microscopes arise from the many different types of interactions that occur when a small probe is scanned over and interacts with a specimen. These interactions or modes can be recorded or mapped as function of location on the surface to form a characterization map. The three most common types of scanning probe microscopes are atomic force microscopes (AFM), near-field scanning optical microscopes (NSOM or SNOM, scanning near-field optical microscopy), and scanning tunneling microscopes (STM). An atomic force microscope has a fine probe, usually of silicon or silicon nitride, attached to a cantilever; the probe is scanned over the surface of the sample, and the forces that cause an interaction between the probe and the surface of the sample are measured and mapped. A near-field scanning optical microscope is similar to an AFM but its probe consists of a light source in an optical fiber covered with a tip that has usually an aperture for the light to pass through. The microscope can capture either transmitted or reflected light to measure very localized optical properties of the surface, commonly of a biological specimen. Scanning tunneling microscopes have a metal tip with a single apical atom; the tip is attached to a tube through which a current flows. The tip is scanned over the surface of a conductive sample until a tunneling current flows; the current is kept constant by computer movement of the tip and an image is formed by the recorded movements of the tip.

Other Types

Scanning acoustic microscopes use sound waves to measure variations in acoustic impedance. Similar to Sonar in principle, they are used for such jobs as detecting defects in the subsurfaces of materials including those found in integrated circuits. On February 4, 2013, Australian engineers built a "quantum microscope" which provides unparalleled precision.

Mobile Apps

Mobile app microscopes can optionally be used as optical microscope when the flashlight is activated. However, mobile app microscopes are harder to use due to visual noise, are often limited to 40x, and the resolution limits of the camera lens itself.

3.2 THE ROLE OF MICROSCOPY IN FORENSIC SCIENCE

Microscopy plays an integral role in forensic investigations by enabling the examination of minute pieces of evidence that are often invisible to the naked eye.

Forensic chemists and scientists utilize microscopes to magnify small particles or materials collected from crime scenes, helping to establish connections between suspects, victims, and locations. Microscopy is particularly valuable in the analysis of trace evidence, which can include substances such as hair, fibers, soil, and glass fragments, among others.

Forensic chemists use various types of microscopes, depending on the nature of the material being examined and the level of detail required.

The information gained through microscopic analysis often provides crucial evidence that can support or refute hypotheses about a crime, ultimately aiding in the identification of suspects and the reconstruction of criminal events.

3.3 TYPES OF MICROSCOPES USED IN FORENSIC CHEMISTRY

There are several types of microscopes used in forensic chemistry, each suited to different types of evidence and levels of analysis. The most common types include the light microscope, comparison microscope, stereomicroscope, polarizing microscope, and scanning electron microscope (SEM).

Each of these devices has unique capabilities that make them suitable for specific forensic applications.

3.3.1 Light Microscope

The light microscope, also known as the optical microscope, is the most basic and widely used type of microscope in forensic chemistry. It uses visible light to magnify small objects, allowing forensic scientists to examine various forms of trace evidence. Light microscopes are useful for analyzing biological materials such as blood, hair, and fibers, as well as non-biological materials such as soil particles or paint chips.

Light microscopy is often used as a preliminary step in forensic analysis, as it provides an overall view of the material being examined. In some cases, forensic chemists may use stains or dyes to enhance the contrast of the specimen, making certain features more visible under the microscope.

3.3.2 Comparison Microscope

The comparison microscope is a vital tool in forensic science, especially in the field of firearms and toolmark analysis. It allows the simultaneous viewing of two separate objects side by side, which is essential for comparing evidence such as bullets, cartridge cases, or tool marks. The ability to observe two objects at the same time under identical conditions helps forensic scientists determine whether the evidence comes from the same source.

In ballistics, for instance, the comparison microscope is used to match markings on a recovered bullet with test bullets fired from a suspect's firearm. Toolmark comparisons are also common, where the tool marks left on an object at the crime scene are compared with test marks created using a suspected tool.

3.3.3 Stereomicroscope

The stereomicroscope, also known as the dissecting microscope, provides a three-dimensional view of the object being examined. This type of microscope is especially useful for examining the surface features and textures of larger objects or particles. It has a lower magnification than other microscopes but offers a wider field of view, which allows forensic chemists to manipulate and handle evidence under the microscope.

Stereomicroscopes are commonly used to examine fibers, hairs, and paint chips, as well as fragments of glass or metal. For example, forensic chemists might use a stereomicroscope to analyze the cut end of a fiber or hair to determine if it was cut, torn, or burned. It is also valuable in the preliminary examination of evidence before more detailed analyses are performed.

3.3.4 Polarizing Microscope

The polarizing microscope is an advanced type of light microscope that uses polarized light to examine materials that are optically active, meaning they change the direction of light passing through them. This type of microscope is particularly useful in the examination of crystalline substances, minerals, and fibers. In forensic chemistry, polarizing microscopes are used to analyze drug samples, identify minerals in soil, and examine synthetic fibers. One of the primary uses of the polarizing microscope is in the identification of different types of synthetic fibers. Forensic chemists can examine the optical properties of fibers, such as birefringence, to determine their composition. This type of analysis can help establish links between crime scenes, victims, and suspects when fibers from clothing or other materials are transferred during criminal activity.

3.3.5 Scanning Electron Microscope (SEM)

The scanning electron microscope (SEM) is one of the most powerful microscopes available in forensic chemistry. It uses a focused beam of electrons rather than light to create highly detailed, three-dimensional images of the surface of a sample. SEM is particularly useful for analyzing small particles and trace evidence, as it provides a much higher magnification and resolution than light microscopes.

SEM is often used in the analysis of gunshot residue (GSR), a key piece of evidence in shooting investigations. When a firearm is discharged, microscopic particles of gunshot residue are expelled from the weapon. By using SEM, forensic chemists can examine these particles and determine their composition, which can help link a suspect to the use of a firearm. SEM is also used to examine other types of trace evidence, such as fibers, glass fragments, and biological materials.

3.4 MICROSCOPIC ANALYSIS OF COMMON TYPES OF FORENSIC EVIDENCE

Microscopic analysis is applied to various types of forensic evidence, each requiring specific techniques and expertise. Some of the most common forms of evidence examined under the microscope in forensic chemistry include hair, fibers, glass, and gunshot residue.

3.4.1 Hair Analysis

Hair is one of the most frequently encountered forms of trace evidence in forensic investigations. Under the microscope, forensic chemists can analyze hair samples to determine characteristics such as color, texture, and structure. The analysis can reveal whether the hair came from a human or an animal, and in some cases, it can help identify the racial background or even the specific individual to whom the hair belongs.

Hair analysis typically begins with light microscopy to examine the overall appearance of the hair. Forensic scientists may then use comparison microscopy to match a hair sample found at the crime scene to a known sample from a suspect. In some cases, more advanced techniques such as SEM are used to examine the cuticle, cortex, and medulla of the hair in greater detail.

3.4.2 Fiber Analysis

Fibers are another common type of trace evidence examined in forensic investigations, particularly in cases involving physical contact between a suspect and a victim or crime scene. Fibers can be transferred between clothing, carpets, upholstery, and other materials, providing valuable evidence of a connection between individuals or locations.

Under the microscope, forensic chemists can analyze the physical and optical properties of fibers to determine their origin. Techniques such as polarizing microscopy and SEM are often used to examine fibers in detail, allowing forensic scientists to distinguish between natural and synthetic fibers and even identify specific manufacturers.

3.4.3 Glass Fragment Analysis

Glass is a common material found at crime scenes, particularly in cases involving break-ins or traffic accidents. Forensic scientists use microscopy to analyze glass fragments, which can provide information about the source of the glass and its role in the crime.

By examining the refractive index and surface features of glass fragments, forensic chemists can link glass found at the crime scene to a specific window, bottle, or other glass objects.

SEM is particularly useful in glass analysis, as it allows forensic scientists to examine the surface texture of glass fractures, helping to reconstruct the events leading to the breakage. Microscopy can also reveal traces of paint or other materials on glass fragments, which can further assist in linking the evidence to a suspect or location.

3.4.4 Gunshot Residue (GSR) Analysis

Gunshot residue analysis is a critical part of firearms investigations. GSR consists of microscopic particles expelled from a firearm when it is discharged, and it can be found on the shooter's hands, clothing, or nearby surfaces. Forensic chemists use SEM to analyze GSR particles, which provides detailed images of their morphology and elemental composition.

By examining the unique characteristics of GSR particles, forensic scientists can determine whether an individual was in close proximity to a discharged firearm. This type of analysis is particularly important in cases involving shootings, as it can help establish the involvement of a suspect.

3.5 THE IMPACT OF MICROSCOPY IN FORENSIC CHEMISTRY

The development and application of advanced microscopy techniques have significantly enhanced the field of forensic chemistry, allowing for more detailed and accurate analyses of evidence.

The ability to examine trace evidence at a microscopic level has led to the resolution of countless criminal cases, providing investigators with the tools they need to solve complex crimes.

Microscopic analysis not only helps in identifying suspects but also plays a crucial role in reconstructing the events of a crime. By analyzing fibers, glass, gunshot residue, and other materials, forensic chemists can piece together the interactions between individuals and objects, helping to establish timelines and connections that would otherwise be impossible to discern.

3.6 CONCLUSION

The microscope is an indispensable tool in forensic chemistry, providing the means to examine trace evidence with a level of precision and detail that is critical to criminal investigations.

From light microscopy to advanced techniques such as SEM, microscopy enables forensic scientists to analyze a wide range of materials, from hair and fibers to glass fragments and gunshot residue. Through the application of these techniques, forensic chemists can uncover valuable evidence that plays a pivotal role in solving crimes and ensuring that justice is served.

CHAPTER 4
LIGHT SPECTROSCOPY

4.1 UNDERSTANDING THE LIGHT SPECTROSCOPY

Light spectroscopy is a powerful analytical tool in forensic chemistry that allows scientists to identify and analyze the composition of substances by examining how they interact with light. This non-destructive technique provides forensic chemists with the ability to detect and quantify trace amounts of materials found at crime scenes, such as drugs, explosives, or toxins. The versatility and accuracy of light spectroscopy make it an essential method for uncovering evidence that may not be detectable through other means.

4.2 THE BASICS OF LIGHT SPECTROSCOPY

Spectroscopy is based on the principle that different materials absorb, emit, or reflect light at specific wavelengths, depending on their chemical structure. By analyzing these interactions, scientists can determine the composition of a sample. In forensic chemistry, light spectroscopy is particularly valuable because it is a rapid, precise, and non-destructive method for analyzing small or trace quantities of substances.

The key concept behind light spectroscopy is the electromagnetic spectrum, which encompasses a wide range of wavelengths, from gamma rays to radio waves. Different regions of the spectrum are used for various types of spectroscopy, such as ultraviolet-visible (UV-Vis) spectroscopy, infrared (IR) spectroscopy, and Raman spectroscopy. Each of these techniques has distinct advantages and applications in forensic analysis.

4.3 TYPES OF LIGHT SPECTROSCOPY USED IN FORENSIC CHEMISTRY

Several types of light spectroscopy are commonly employed in forensic chemistry, depending on the nature of the evidence and the specific information required. The following are the most frequently used techniques in forensic investigations:

4.3.1 Ultraviolet-Visible (UV-Vis) Spectroscopy

UV-Vis spectroscopy measures the absorption or transmission of ultraviolet and visible light by a sample. When light passes through a substance, certain wavelengths are absorbed by the molecules, causing electronic transitions between energy levels. The resulting spectrum can be used to identify and quantify specific compounds, such as drugs or poisons, based on their unique absorption patterns.

In forensic chemistry, UV-Vis spectroscopy is commonly used to analyze colorants in fibers, paints, and inks. It is also applied to detect trace amounts of narcotics, as many drugs have characteristic absorption spectra in the UV region. The technique is highly sensitive and can provide information about the concentration of substances in a sample, making it useful for quantitative analysis.

4.3.2 Infrared (IR) Spectroscopy

Infrared spectroscopy involves the interaction of infrared light with a sample, causing vibrations in the molecular bonds. Each molecule has a unique vibrational pattern, producing a distinct infrared spectrum that acts as a molecular fingerprint. By comparing the spectrum of an unknown substance to reference spectra of known compounds, forensic chemists can identify the chemical composition of the sample.

IR spectroscopy is particularly useful for analyzing organic compounds, such as drugs, explosives, and polymers. In forensic investigations, it is often used to examine substances like drugs of abuse, fibers, and paint chips. For example, IR spectroscopy can determine the type of binder or pigment used in a paint sample, which can help link a suspect to a crime scene. The non-destructive nature of IR spectroscopy allows forensic scientists to preserve the evidence for further analysis.

4.3.3 Raman Spectroscopy

Raman spectroscopy is another technique based on the interaction of light with molecular vibrations. However, instead of measuring absorption, Raman spectroscopy detects scattered light. When light interacts with a molecule, a small portion of the light is scattered at different wavelengths, depending on the vibrational modes of the molecule.

This scattered light forms a spectrum that provides information about the molecular structure of the sample.

Raman spectroscopy is particularly valuable in forensic chemistry because it can be used to analyze a wide variety of materials, including drugs, explosives, and biological samples. One of its key advantages is that it requires minimal sample preparation and can be performed directly on solid or liquid samples. This makes it ideal for analyzing evidence that cannot be easily dissolved or altered, such as trace materials on fabrics or surfaces. Raman spectroscopy is also highly selective, allowing forensic chemists to distinguish between closely related compounds.

4.3.4 Atomic Absorption Spectroscopy (AAS)

Atomic absorption spectroscopy (AAS) is a technique used to analyze the presence of metal elements in a sample. When a sample is vaporized in a flame or furnace, the atoms absorb light at specific wavelengths. The amount of light absorbed corresponds to the concentration of the element in the sample, allowing for quantitative analysis.

AAS is particularly useful in forensic chemistry for detecting trace metals in toxicology investigations, environmental crime scenes, or cases involving heavy metal poisoning.

For example, AAS can be used to determine the levels of lead, mercury, or arsenic in biological samples such as blood or tissue. The technique is highly sensitive and can detect elements at very low concentrations, making it a valuable tool for analyzing evidence in criminal cases.

4.3.5 Fourier Transform Infrared (FTIR) Spectroscopy

Fourier Transform Infrared (FTIR) spectroscopy is a variation of infrared spectroscopy that uses a mathematical algorithm called Fourier transform to convert raw data into a readable spectrum. FTIR spectroscopy allows for the rapid collection of high-resolution spectra, making it a widely used technique in forensic laboratories.

FTIR spectroscopy is applied in the identification of a wide range of substances, including drugs, explosives, fibers, and polymers. Its ability to produce highly detailed spectra allows forensic chemists to differentiate between materials with similar chemical compositions.

The technique is also non-destructive, meaning that the sample remains intact for further testing.

4.4 APPLICATIONS OF LIGHT SPECTROSCOPY IN FORENSIC CHEMISTRY

Light spectroscopy has a wide range of applications in forensic chemistry, from identifying substances to linking suspects to crime scenes. Below are some key areas where spectroscopy plays a vital role in forensic investigations.

4.4.1 Drug Analysis

One of the most common applications of light spectroscopy in forensic chemistry is the analysis of controlled substances. Forensic chemists use techniques like UV-Vis and IR spectroscopy to identify drugs in seized materials or biological samples.

By comparing the absorption or vibrational spectra of an unknown sample to reference data, they can determine the identity and concentration of the drug.

Spectroscopic methods are particularly useful for detecting drugs in trace amounts or mixed with other substances. For example, Raman spectroscopy can be used to analyze street drugs that may be adulterated with cutting agents, helping to establish the purity and composition of the seized material.

4.4.2 Explosives Detection

The detection and analysis of explosives are critical in forensic investigations, particularly in cases involving terrorism or bombings. Infrared and Raman spectroscopy are widely used to identify explosive materials and their residues. These techniques allow forensic chemists to analyze bomb fragments, soil samples, and other materials for traces of explosive compounds.

Spectroscopy can also help determine the chemical composition of improvised explosive devices (IEDs), providing valuable information about the materials used in their construction. This information can be used to trace the source of the explosives and link them to suspects or terrorist organizations.

4.4.3 Fiber and Paint Analysis

Fibers and paints are common forms of trace evidence found at crime scenes, especially in cases involving physical contact or vehicular accidents. Light spectroscopy techniques like UV-Vis and IR spectroscopy are used to analyze the chemical composition of fibers and paints, helping to match evidence to specific sources.

For example, forensic chemists can use IR spectroscopy to analyze the polymer composition of synthetic fibers, which can help identify the manufacturer or type of fabric.

Similarly, spectroscopic analysis of paint chips can reveal the pigments and binders used in the paint, allowing investigators to link the evidence to a particular vehicle or object.

4.4.4 Toxicology and Poison Detection

In toxicology cases, light spectroscopy is used to detect and quantify toxic substances in biological samples such as blood, urine, or tissue. Atomic absorption spectroscopy (AAS) is particularly useful for detecting trace metals like lead, arsenic, or mercury, which are common in cases of poisoning or environmental contamination.

Spectroscopic techniques can also be used to analyze volatile organic compounds (VOCs) and other toxic chemicals. For example, FTIR spectroscopy can detect the presence of carbon monoxide in blood samples, providing evidence in cases of carbon monoxide poisoning.

4.4.5 Gunshot Residue (GSR) Analysis

Gunshot residue (GSR) is a valuable form of evidence in shooting investigations. Light spectroscopy techniques, particularly atomic absorption spectroscopy and Raman spectroscopy, are used to analyze GSR particles. By examining the chemical composition of the residue, forensic chemists can determine whether an individual has recently fired a gun.

Spectroscopic analysis of GSR can also provide information about the type of ammunition used and the distance between the shooter and the victim. This information is crucial for reconstructing the events of a shooting and linking suspects to the use of firearms.

4.5 ADVANTAGES OF LIGHT SPECTROSCOPY IN FORENSIC CHEMISTRY

Light spectroscopy offers several advantages in forensic chemistry, making it one of the most reliable and widely used analytical techniques. Some of the key benefits include:

- **Non-Destructive Analysis**: Spectroscopic techniques do not alter or destroy the sample, allowing forensic scientists to preserve evidence for further analysis.
- **High Sensitivity**: Light spectroscopy can detect trace amounts of substances, making it ideal for analyzing small or diluted samples.
- **Rapid and Accurate**: Spectroscopic methods provide quick and precise results, which are crucial in time-sensitive criminal investigations.
- **Versatile**: Spectroscopy can be used to analyze a wide variety of materials, including biological samples, fibers, paints, drugs, and explosives.
- **Quantitative and Qualitative Analysis**: Light spectroscopy can not only identify substances but also quantify their concentration in a sample.

4.6 CONCLUSION

Light spectroscopy plays a critical role in forensic chemistry, providing a powerful tool for the analysis and identification of materials in criminal investigations. Through techniques such as UV-Vis, IR, Raman, and atomic absorption spectroscopy, forensic chemists can examine trace evidence with high sensitivity and precision. Whether analyzing drugs, explosives, fibers, or toxic substances, light spectroscopy enables the accurate and rapid detection of materials, helping to solve crimes and support the justice system. Its non-destructive nature and versatility make it an indispensable technique in the modern forensic laboratory.

CHAPTER 5
ADVANCED SPECTROSCOPY

5.1 UNDERSTANDING THE ADVANCED SPECTROSCOPY

Forensic chemistry has witnessed significant advancements over the past few decades, with the introduction of sophisticated analytical techniques aimed at enhancing the accuracy and reliability of evidence analysis. Among these methods, advanced spectroscopy has emerged as one of the most powerful tools for forensic chemists. Spectroscopic techniques allow for the precise identification and quantification of substances, even at trace levels, enabling forensic experts to gather critical information from crime scene samples.

5.2 OVERVIEW OF SPECTROSCOPY IN FORENSIC CHEMISTRY

Spectroscopy is based on the principle that different substances absorb, emit, or reflect electromagnetic radiation in unique ways, depending on their molecular structures. This interaction between matter and light can be measured and analyzed to identify and characterize materials. In forensic chemistry, spectroscopy is used to analyze a wide range of evidence, from drugs and explosives to paints, fibers, and biological materials.

While traditional spectroscopic methods such as ultraviolet-visible (UV-Vis) and infrared (IR) spectroscopy have long been used in forensic investigations, advanced spectroscopic techniques offer greater sensitivity, resolution, and specificity. These methods have expanded the forensic chemist's toolkit, providing the ability to analyze increasingly complex and minute samples.

5.3 TYPES OF ADVANCED SPECTROSCOPIC TECHNIQUES IN FORENSIC CHEMISTRY

Several advanced spectroscopic techniques are now widely used in forensic laboratories. Each method has its unique strengths and applications, depending on the type of evidence being analyzed and the information required. The following sections discuss some of the most prominent advanced spectroscopic techniques used in forensic chemistry.

5.3.1 Fourier Transform Infrared (FTIR) Spectroscopy

Fourier-transform infrared spectroscopy (FTIR) is a technique used to obtain an infrared spectrum of absorption or emission of a solid, liquid, or gas. An FTIR spectrometer simultaneously collects high-resolution spectral data over a wide spectral range. This confers a significant advantage over a dispersive spectrometer, which measures intensity over a narrow range of wavelengths at a time. The term *Fourier-transform infrared spectroscopy* originates from the fact that a Fourier transform (a mathematical process) is required to convert the raw data into the actual spectrum.

The goal of absorption spectroscopy techniques (FTIR, ultraviolet-visible ("UV-vis") spectroscopy, etc.) is to measure how much light a sample absorbs at each wavelength. The most straightforward way to do this, the "dispersive spectroscopy" technique, is to shine a monochromatic light beam at a sample, measure how much of the light is absorbed, and repeat for each different wavelength. (This is how some UV–vis spectrometers work, for example.)

Fourier-transform spectroscopy is a less intuitive way to obtain the same information. Rather than shining a monochromatic beam of light (a beam composed of only a single wavelength) at the sample, this technique shines a beam containing many frequencies of light at once and measures how much of that beam is absorbed by the sample. Next, the beam is modified to contain a different combination of frequencies, giving a second data point. This process is rapidly repeated many times over a short time span. Afterwards, a computer takes all this data and works backward to infer what the absorption is at each wavelength.

The beam described above is generated by starting with a broadband light source—one containing the full spectrum of wavelengths to be measured. The light shines into a Michelson interferometer—a certain configuration of mirrors, one of which is moved by a motor. As this mirror moves, each wavelength of light in the beam is periodically blocked, transmitted, blocked, transmitted, by the interferometer, due to wave interference. Different wavelengths are modulated at different rates, so that at each moment or mirror position the beam coming out of the interferometer has a different spectrum.

As mentioned, computer processing is required to turn the raw data (light absorption for each mirror position) into the desired result (light absorption for each wavelength). The processing required turns out to be a common algorithm called the Fourier transform. The Fourier transform converts one domain (in this case displacement of the mirror in cm) into its inverse domain (wavenumbers in cm^{-1}). The raw data is called an "interferogram".

History

The first low-cost spectrophotometer capable of recording an infrared spectrum was the Perkin-Elmer Infracord produced in 1957. This instrument covered the wavelength range from 2.5 μm to 15 μm (wavenumber range 4,000 cm^{-1} to 660 cm^{-1}).

The lower wavelength limit was chosen to encompass the highest known vibration frequency due to a fundamental molecular vibration. The upper limit was imposed by the fact that the dispersing element was a prism made from a single crystal of rock-salt (sodium chloride), which becomes opaque at wavelengths longer than about 15 μm; this spectral region became known as the rock-salt region. Later instruments used potassium bromide prisms to extend the range to 25 μm (400 cm^{-1}) and caesium iodide 50 μm (200 cm^{-1}). The region beyond 50 μm (200 cm^{-1}) became known as the far-infrared region; at very long wavelengths it merges into the microwave region. Measurements in the far infrared needed the development of accurately ruled diffraction gratings to replace the prisms as dispersing elements, since salt crystals are opaque in this region. More sensitive detectors than the bolometer were required because of the low energy of the radiation. One such was the Golay detector. An additional issue is the need to exclude atmospheric water vapour because water vapour has an intense pure rotational spectrum in this region. Far-infrared spectrophotometers were cumbersome, slow and expensive. The advantages of the Michelson interferometer were well-known, but considerable technical difficulties had to be overcome before a commercial instrument could be built. Also an electronic computer was needed to perform the required Fourier transform, and this only became practicable with the advent of minicomputers, such as the PDP-8, which became available in 1965. Digilab pioneered the world's first commercial FTIR spectrometer (Model FTS-14) in 1969. Digilab FTIRs are now a part of Agilent technologies's molecular product line after Agilent acquired spectroscopy business from Varian.

Michelson Interferometer

In a Michelson interferometer adapted for FTIR, light from the polychromatic infrared source, approximately a black-body radiator, is collimated and directed to a beam splitter. Ideally 50% of the light is refracted towards the fixed mirror and 50% is transmitted towards the moving mirror. Light is reflected from the two mirrors back to the beam splitter and some fraction of the original light passes into the sample compartment. There, the light is focused on the sample. On leaving the sample compartment the light is refocused on to the detector. The difference in optical path length between the two arms to the interferometer is known as the *retardation* or *optical path difference* (OPD). An interferogram is obtained by varying the retardation and recording the signal from the detector for various values of the retardation. The form of the interferogram when no sample is present depends on factors such as the variation of source intensity and splitter efficiency with wavelength. This results in a maximum at zero retardation, when there is constructive interference at all wavelengths, followed by series of "wiggles". The position of zero retardation is determined accurately by finding the point of maximum intensity in the interferogram. When a sample is present the background interferogram is modulated by the presence of absorption bands in the sample.

Commercial spectrometers use Michelson interferometers with a variety of scanning mechanisms to generate the path difference. Common to all these arrangements is the need to ensure that the two beams recombine exactly as the system scans. The simplest systems have a plane mirror that moves linearly to vary the path of one beam. In this arrangement the moving mirror must not tilt or wobble as this would affect how the beams overlap as they recombine. Some systems incorporate a compensating mechanism that automatically adjusts the orientation of one mirror to maintain the alignment. Arrangements that avoid this problem include using cube corner reflectors instead of plane mirrors as these have the property of returning any incident beam in a parallel direction regardless of orientation.

Systems where the path difference is generated by a rotary movement have proved very successful. One common system incorporates a pair of parallel mirrors in one beam that can be rotated to vary the path without displacing the returning beam. Another is the double pendulum design where the path in one arm of the interferometer increases as the path in the other decreases.

A quite different approach involves moving a wedge of an IR-transparent material such as KBr into one of the beams. Increasing the thickness of KBr in the beam increases the optical path because the refractive index is higher than that of air. One limitation of this approach is that the variation of refractive index over the wavelength range limits the accuracy of the wavelength calibration.

Measuring and Processing the Interferogram

The interferogram has to be measured from zero path difference to a maximum length that depends on the resolution required. In practice the scan can be on either side of zero resulting in a double-sided interferogram. Mechanical design limitations may mean that for the highest resolution the scan runs to the maximum OPD on one side of zero only.

The interferogram is converted to a spectrum by Fourier transformation. This requires it to be stored in digital form as a series of values at equal intervals of the path difference between the two beams. To measure the path difference a laser beam is sent through the interferometer, generating a sinusoidal signal where the separation between successive maxima is equal to the wavelength of the laser (typically a 633 nm HeNe laser is used). This can trigger an analog-to-digital converter to measure the IR signal each time the laser signal passes through zero. Alternatively, the laser and IR signals can be measured synchronously at smaller intervals with the IR signal at points corresponding to the laser signal zero crossing being determined by interpolation. This approach allows the use of analog-to-digital converters that are more accurate and precise than converters that can be triggered, resulting in lower noise.

The result of Fourier transformation is a spectrum of the signal at a series of discrete wavelengths. The range of wavelengths that can be used in the calculation is limited by the separation of the data points in the interferogram. The shortest wavelength that can be recognized is twice the separation between these data points. For example, with one point per wavelength of a HeNe reference laser at 0.633 μm (15800 cm^{-1}) the shortest wavelength would be 1.266 μm (7900 cm^{-1}). Because of aliasing, any energy at shorter wavelengths would be interpreted as coming from longer wavelengths and so has to be minimized optically or electronically. The spectral resolution, i.e. the separation between wavelengths that can be distinguished, is determined by the maximum OPD. The wavelengths used in calculating the Fourier transform are such that an exact number of wavelengths fit into the length of the interferogram from zero to the maximum OPD as this makes their contributions orthogonal. This results in a spectrum with points separated by equal frequency intervals.

For a maximum path difference *d* adjacent wavelengths λ_1 and λ_2 will have *n* and *(n+1)* cycles, respectively, in the interferogram. The corresponding frequencies are ν_1 and ν_2:

$d = n\lambda_1$	and $d = (n+1)\lambda_2$
$\lambda_1 = d/n$	and $\lambda_2 = d/(n+1)$
$\nu_1 = 1/\lambda_1$	and $\nu_2 = 1/\lambda_2$
$\nu_1 = n/d$	and $\nu_2 = (n+1)/d$
$\nu_2 - \nu_1 = 1/d$	

The separation is the inverse of the maximum OPD. For example, a maximum OPD of 2 cm results in a separation of 0.5 cm^{-1}. This is the spectral resolution in the sense that the value at one point is independent of the values at adjacent points. Most instruments can be operated at different resolutions by choosing different OPD's. Instruments for routine analyses typically have a best resolution of around 0.5 cm^{-1}, while spectrometers have been built with resolutions as high as 0.001 cm^{-1}, corresponding to a maximum OPD of 10 m. The point in the interferogram corresponding to zero path difference has to be identified, commonly by assuming it is where the maximum signal occurs. This so-called centerburst is not always symmetrical in real world spectrometers so a phase correction may have to be calculated. The interferogram signal decays as the path difference increases, the rate of decay being inversely related to the width of features in the spectrum. If the OPD is not large enough to allow the interferogram signal to decay to a negligible level there will be unwanted oscillations or sidelobes associated with the features in the resulting spectrum. To reduce these sidelobes the interferogram is usually multiplied by a function that approaches zero at the maximum OPD. This so-called apodization reduces the amplitude of any sidelobes and also the noise level at the expense of some reduction in resolution.

For rapid calculation the number of points in the interferogram has to equal a power of two. A string of zeroes may be added to the measured interferogram to achieve this. More zeroes may be added in a process called zero filling to improve the appearance of the final spectrum although there is no improvement in resolution. Alternatively, interpolation after the Fourier transform gives a similar result.

Advantages

There are three principal advantages for an FT spectrometer compared to a scanning (dispersive) spectrometer.

- **The Multiplex or Fellgett's Advantage:** This arises from the fact that information from all wavelengths is collected simultaneously. It results in a higher signal-to-noise ratio for a given scan-time for observations limited by a fixed detector noise contribution (typically in the thermal infrared spectral region where a photodetector is limited by generation-recombination noise). For a spectrum with m resolution elements, this increase is equal to the square root of m. Alternatively, it allows a shorter scan-time for a given resolution. In practice multiple scans are often averaged, increasing the signal-to-noise ratio by the square root of the number of scans.
- **The Throughput or Jacquinot's Advantage:** This results from the fact that in a dispersive instrument, the monochromator has entrance and exit slits which restrict the amount of light that passes through it. The interferometer throughput is determined only by the diameter of the collimated beam coming from the source. Although no slits are needed, FTIR spectrometers do require an aperture to restrict the convergence of the collimated beam in the interferometer. This is because convergent rays are modulated at different frequencies as the path difference is varied. Such an aperture is called a Jacquinot stop. For a given resolution and wavelength this circular aperture allows more light through than a slit, resulting in a higher signal-to-noise ratio.
- **The Wavelength Accuracy or Connes' Advantage:** The wavelength scale is calibrated by a laser beam of known wavelength that passes through the interferometer. This is much more stable and accurate than in dispersive instruments where the scale depends on the mechanical

movement of diffraction gratings. In practice, the accuracy is limited by the divergence of the beam in the interferometer which depends on the resolution.

Another minor advantage is less sensitivity to stray light, that is radiation of one wavelength appearing at another wavelength in the spectrum. In dispersive instruments, this is the result of imperfections in the diffraction gratings and accidental reflections. In FT instruments there is no direct equivalent as the apparent wavelength is determined by the modulation frequency in the interferometer.

Resolution

The interferogram belongs in the length dimension. Fourier transform (FT) inverts the dimension, so the FT of the interferogram belongs in the reciprocal length dimension([L^{-1}]), that is the dimension of wavenumber. The spectral resolution in cm^{-1} is equal to the reciprocal of the maximal retardation in cm. Thus a 4 cm^{-1} resolution will be obtained if the maximal retardation is 0.25 cm; this is typical of the cheaper FTIR instruments. Much higher resolution can be obtained by increasing the maximal retardation. This is not easy, as the moving mirror must travel in a near-perfect straight line. The use of corner-cube mirrors in place of the flat mirrors is helpful, as an outgoing ray from a corner-cube mirror is parallel to the incoming ray, regardless of the orientation of the mirror about axes perpendicular to the axis of the light beam.

A spectrometer with 0.001 cm^{-1} resolution is now available commercially. The throughput advantage is important for high-resolution FTIR, as the monochromator in a dispersive instrument with the same resolution would have very narrow entrance and exit slits.

In 1966 Janine Connes measured the temperature of the atmosphere of Venus by recording the vibration-rotation spectrum of Venusian CO_2 at 0.1 cm^{-1} resolution. Michelson himself attempted to resolve the hydrogen H_α emission band in the spectrum of a hydrogen atom into its two components by using his interferometer.

Components

IR Sources

FTIR spectrometers are mostly used for measurements in the mid and near IR regions. For the mid-IR region, 2–25 μm (5,000–400 cm^{-1}), the most common source is a silicon carbide (SiC) element heated to about 1,200 K (930 °C; 1,700 °F) (Globar). The output is similar to a blackbody. Shorter wavelengths of the near-IR, 1–2.5 μm (10,000–4,000 cm^{-1}), require a higher temperature source, typically a tungsten-halogen lamp. The long wavelength output of these is limited to about 5 μm (2,000 cm^{-1}) by the absorption of the quartz envelope. For the far-IR, especially at wavelengths beyond 50 μm (200 cm^{-1}) a mercury discharge lamp gives higher output than a thermal source.

Detectors

Far-IR spectrometers commonly use pyroelectric detectors that respond to changes in temperature as the intensity of IR radiation falling on them varies. The sensitive elements in these detectors are either deuterated triglycine sulfate (DTGS) or lithium tantalate ($LiTaO_3$). These detectors operate at ambient temperatures and provide adequate sensitivity for most routine applications. To achieve the best sensitivity the time for a scan is typically a few seconds. Cooled photoelectric detectors are employed for situations requiring higher sensitivity or faster response. Liquid nitrogen cooled mercury cadmium telluride (MCT) detectors are the most widely used in the mid-IR. With these detectors an interferogram can be measured in as little as 10 milliseconds. Uncooled indium gallium arsenide photodiodes or DTGS are the usual choices in near-IR systems. Very sensitive liquid-helium-cooled silicon or germanium bolometers are used in the far-IR where both sources and beamsplitters are inefficient.

Beam Splitter

An ideal beam-splitter transmits and reflects 50% of the incident radiation. However, as any material has a limited range of optical transmittance, several beam-splitters may be used interchangeably to cover a wide spectral range.

In a simple Michelson interferometer, one beam passes twice through the beamsplitter but the other passes through only once. To correct for this, an additional compensator plate of equal thickness is incorporated.

For the mid-IR region, the beamsplitter is usually made of KBr with a germanium-based coating that makes it semi-reflective. KBr absorbs strongly at wavelengths beyond 25 μm (400 cm^{-1}), so CsI or KRS-5 are sometimes used to extend the range to about 50 μm (200 cm^{-1}). ZnSe is an alternative where moisture vapour can be a problem, but is limited to about 20 μm (500 cm^{-1}).

CaF_2 is the usual material for the near-IR, being both harder and less sensitive to moisture than KBr, but cannot be used beyond about 8 μm (1,200 cm^{-1}). Far-IR beamsplitters are mostly based on polymer films, and cover a limited wavelength range.

Attenuated Total Reflectance

Attenuated total reflectance (ATR) is one accessory of FTIR spectrophotometer to measure surface properties of solid or thin film samples rather than their bulk properties. Generally, ATR has a penetration depth of around 1 or 2 micrometers depending on sample conditions.

Fourier Transform

The interferogram in practice consists of a set of intensities measured for discrete values of retardation. The difference between successive retardation values is constant. Thus, a discrete Fourier transform is needed. The fast Fourier transform (FFT) algorithm is used.

Spectral Range

Far-Infrared

The first FTIR spectrometers were developed for far-infrared range. The reason for this has to do with the mechanical tolerance needed for good optical performance, which is related to the wavelength of the light being used. For the relatively long wavelengths of the far infrared, ~10 μm tolerances are adequate, whereas for the rock-salt region tolerances have to be better than 1 μm. A typical instrument was the cube interferometer developed at the NPL and marketed by Grubb Parsons. It used a stepper motor to drive the moving mirror, recording the detector response after each step was completed.

Mid-Infrared

With the advent of cheap microcomputers it became possible to have a computer dedicated to controlling the spectrometer, collecting the data, doing the Fourier transform and presenting the spectrum. This provided the impetus for the development of FTIR spectrometers for the rock-salt region. The problems of manufacturing ultra-high precision optical and mechanical components had to be solved. A wide range of instruments are now available commercially. Although instrument design has become more sophisticated, the basic principles remain the same. Nowadays, the moving mirror of the interferometer moves at a constant velocity, and sampling of the interferogram is triggered by finding zero-crossings in the fringes of a secondary interferometer lit by a helium–neon laser. In modern FTIR systems the constant mirror velocity is not strictly required, as long as the laser fringes and the original interferogram are recorded simultaneously with higher sampling rate and then re-interpolated on a constant grid, as pioneered by James W. Brault. This confers very high wavenumber accuracy on the resulting infrared spectrum and avoids wavenumber calibration errors.

Near-Infrared

The near-infrared region spans the wavelength range between the rock-salt region and the start of the visible region at about 750 nm. Overtones of fundamental vibrations can be observed in this region. It is used mainly in industrial applications such as process control and chemical imaging.

Applications

FTIR can be used in all applications where a dispersive spectrometer was used in the past. In addition, the improved sensitivity and speed have opened up new areas of application. Spectra can be measured in situations where very little energy reaches the detector. Fourier transform infrared spectroscopy is used in geology, chemistry, materials, botany and biology research fields.

Nano and Biological Materials

FTIR is also used to investigate various nanomaterials and proteins in hydrophobic membrane environments.

Studies show the ability of FTIR to directly determine the polarity at a given site along the backbone of a transmembrane protein. The bond features involved with various organic and inorganic nanomaterials and their quantitative analysis can be done with the help of FTIR.

Microscopy and Imaging

An infrared microscope allows samples to be observed and spectra measured from regions as small as 5 microns across. Images can be generated by combining a microscope with linear or 2-D array detectors. The spatial resolution can approach 5 microns with tens of thousands of pixels. The images contain a spectrum for each pixel and can be viewed as maps showing the intensity at any wavelength or combination of wavelengths. This allows the distribution of different chemical species within the sample to be seen. This technique has been applied in various biological applications including the analysis of tissue sections as an alternative to conventional histopathology, examining the homogeneity of pharmaceutical tablets, and for differentiating morphologically-similar pollen grains.

Nanoscale and Spectroscopy below the Diffraction Limit

The spatial resolution of FTIR can be further improved below the micrometer scale by integrating it into scanning near-field optical microscopy platform. The corresponding technique is called nano-FTIR and allows for performing broadband spectroscopy on materials in ultra-small quantities (single viruses and protein complexes) and with 10 to 20 nm spatial resolution.

FTIR as Detector in Chromatography

The speed of FTIR allows spectra to be obtained from compounds as they are separated by a gas chromatograph. However this technique is little used compared to GC-MS (gas chromatography-mass spectrometry) which is more sensitive. The GC-IR method is particularly useful for identifying isomers, which by their nature have identical masses. Liquid chromatography fractions are more difficult because of the solvent present. One notable exception is to measure chain branching as a function of molecular size in polyethylene using gel permeation chromatography, which is possible using chlorinated solvents that have no absorption in the area in question.

TG-IR (Thermogravimetric Analysis-Infrared Spectrometry)

Measuring the gas evolved as a material is heated allows qualitative identification of the species to complement the purely quantitative information provided by measuring the weight loss.

Water Content Determination in Plastics and Composites

FTIR analysis is used to determine water content in fairly thin plastic and composite parts, more commonly in the laboratory setting. Such FTIR methods have long been used for plastics, and became extended for composite materials in 2018, when the method was introduced by Krauklis, Gagani and Echtermeyer. FTIR method uses the maxima of the absorbance band at about 5,200 cm−1 which correlates with the true water content in the material.

5.3.2 Raman Spectroscopy

Raman spectroscopy (named after physicist C. V. Raman) is a spectroscopic technique typically used to determine vibrational modes of molecules, although rotational and other low-frequency modes of systems may also be observed. Raman spectroscopy is commonly used in chemistry to provide a structural fingerprint by which molecules can be identified.

Raman spectroscopy relies upon inelastic scattering of photons, known as Raman scattering. A source of monochromatic light, usually from a laser in the visible, near infrared, or near ultraviolet range is used, although X-rays can also be used. The laser light interacts with molecular vibrations, phonons or other excitations in the system, resulting in the energy of the laser photons being shifted up or down. The shift in energy gives information about the vibrational modes in the system. Infrared spectroscopy typically yields similar yet complementary information.

Typically, a sample is illuminated with a laser beam. Electromagnetic radiation from the illuminated spot is collected with a lens and sent through a monochromator. Elastic scattered radiation at the wavelength corresponding to the laser line (Rayleigh scattering) is filtered out by either a notch filter, edge pass filter, or a band pass filter, while the rest of the collected light is dispersed onto a detector.

Spontaneous Raman scattering is typically very weak; as a result, for many years the main difficulty in collecting Raman spectra was separating the weak inelastically scattered light from the intense Rayleigh scattered laser light (referred to as "laser rejection"). Historically, Raman spectrometers used holographic gratings and multiple dispersion stages to achieve a high degree of laser rejection. In the past, photomultipliers were the detectors of choice for dispersive Raman setups, which resulted in long acquisition times. However, modern instrumentation almost universally employs notch or edge filters for laser rejection. Dispersive single-stage spectrographs (axial transmissive (AT) or Czerny–Turner (CT) monochromators) paired with CCD detectors are most common although Fourier transform (FT) spectrometers are also common for use with NIR lasers.

The name "Raman spectroscopy" typically refers to vibrational Raman using laser wavelengths which are not absorbed by the sample. There are many other variations of Raman spectroscopy including surface-enhanced Raman, resonance Raman, tip-enhanced Raman, polarized Raman, stimulated Raman, transmission Raman, spatially-offset Raman, and hyper Raman.

History

Although the inelastic scattering of light was predicted by Adolf Smekal in 1923, it was not observed in practice until 1928. The Raman effect was named after one of its discoverers, the Indian scientist C. V. Raman, who observed the effect in organic liquids in 1928 together with K. S. Krishnan, and independently by Grigory Landsberg and Leonid Mandelstam in inorganic crystals. Raman won the Nobel Prize in Physics in 1930 for this discovery. The first observation of Raman spectra in gases was in 1929 by Franco Rasetti.

Systematic pioneering theory of the Raman effect was developed by Czechoslovak physicist George Placzek between 1930 and 1934. The mercury arc became the principal light source, first with photographic detection and then with spectrophotometric detection.

In the years following its discovery, Raman spectroscopy was used to provide the first catalog of molecular vibrational frequencies. Typically, the sample was held in a long tube and illuminated along its length with a beam of filtered monochromatic light generated by a gas discharge lamp. The photons that were scattered by the sample were collected through an optical flat at the end of the tube. To maximize the sensitivity, the sample was highly concentrated (1 M or more) and relatively large volumes (5 mL or more) were used.

Theory

The magnitude of the Raman effect correlates with polarizability of the electrons in a molecule. It is a form of inelastic light scattering, where a photon excites the sample. This excitation puts the molecule into a virtual energy state for a short time before the photon is emitted. Inelastic scattering means that the energy of the emitted photon is of either lower or higher energy than the incident photon. After the scattering event, the sample is in a different rotational or vibrational state.

For the total energy of the system to remain constant after the molecule moves to a new rovibronic (rotational–vibrational–electronic) state, the scattered photon shifts to a different energy, and therefore a different frequency. This energy difference is equal to that between the initial and final rovibronic states of the molecule. If the final state is higher in energy than the initial state, the scattered photon will be shifted to a lower frequency (lower energy) so that the total energy remains the same. This shift in frequency is called a Stokes shift, or downshift. If the final state is lower in energy, the scattered photon will be shifted to a higher frequency, which is called an anti-Stokes shift, or upshift.

For a molecule to exhibit a Raman effect, there must be a change in its electric dipole-electric dipole polarizability with respect to the vibrational coordinate corresponding to the rovibronic state. The intensity of the Raman scattering is proportional to this polarizability change. Therefore, the Raman spectrum (scattering intensity as a function of the frequency shifts) depends on the rovibronic states of the molecule.

The Raman effect is based on the interaction between the electron cloud of a sample and the external electric field of the monochromatic light, which can create an induced dipole moment within the molecule based on its polarizability. Because the laser light does not excite the molecule there can be no real transition between energy levels.

The Raman effect should not be confused with emission (fluorescence or phosphorescence), where a molecule in an excited electronic state emits a photon and returns to the ground electronic state, in many cases to a vibrationally excited state on the ground electronic state potential energy surface. Raman scattering also contrasts with infrared (IR) absorption, where the energy of the absorbed photon matches the difference in energy between the initial and final rovibronic states.

The dependence of Raman on the electric dipole-electric dipole polarizability derivative also differs from IR spectroscopy, which depends on the electric dipole moment derivative, the atomic polar tensor (APT). This contrasting feature allows rovibronic transitions that might not be active in IR to be analyzed using Raman spectroscopy, as exemplified by the rule of mutual exclusion in centrosymmetric molecules.

Transitions which have large Raman intensities often have weak IR intensities and vice versa. If a bond is strongly polarized, a small change in its length such as that which occurs during a vibration has only a small resultant effect on polarization. Vibrations involving polar bonds (e.g. C-O, N-O, O-H) are therefore, comparatively weak Raman scatterers.

Such polarized bonds, however, carry their electrical charges during the vibrational motion, (unless neutralized by symmetry factors), and this results in a larger net dipole moment change during the vibration, producing a strong IR absorption band. Conversely, relatively neutral bonds (e.g. C-C, C-H, C=C) suffer large changes in polarizability during a vibration. However, the dipole moment is not similarly affected such that while vibrations involving predominantly this type of bond are strong Raman scatterers, they are weak in the IR. A third vibrational spectroscopy technique, inelastic incoherent neutron scattering (IINS), can be used to determine the frequencies of vibrations in highly symmetric molecules that may be both IR and Raman inactive.

The IINS selection rules, or allowed transitions, differ from those of IR and Raman, so the three techniques are complementary. They all give the same frequency for a given vibrational transition, but the relative intensities provide different information due to the different types of interaction between the molecule and the incoming particles, photons for IR and Raman, and neutrons for IINS.

Raman Shift

Raman shifts are typically reported in wavenumbers, which have units of inverse length, as this value is directly related to energy. In order to convert between spectral wavelength and wavenumbers of shift in the Raman spectrum, the following formula can be used:

$$\Delta\tilde{\nu} = \left(\frac{1}{\lambda_0} - \frac{1}{\lambda_1}\right),$$

Instrumentation

Modern Raman spectroscopy nearly always involves the use of lasers as excitation light sources. Because lasers were not available until more than three decades after the discovery of the effect, Raman and Krishnan used a mercury lamp and photographic plates to record spectra. Early spectra took hours or even days to acquire due to weak light sources, poor sensitivity of the detectors and the weak Raman scattering cross-sections of most materials. Various colored filters and chemical solutions were used to select certain wavelength regions for excitation and detection but the photographic spectra were still dominated by a broad center line corresponding to Rayleigh scattering of the excitation source.

Technological advances have made Raman spectroscopy much more sensitive, particularly since the 1980s. The most common modern detectors are now charge-coupled devices (CCDs). Photodiode arrays and photomultiplier tubes were common prior to the adoption of CCDs. The advent of reliable, stable, inexpensive lasers with narrow bandwidths has also had an impact.

Lasers

Raman spectroscopy requires a light source such as a laser. The resolution of the spectrum relies on the bandwidth of the laser source used. Generally shorter wavelength lasers give stronger Raman scattering due to the ν^4 increase in Raman scattering cross-sections, but issues with sample degradation or fluorescence may result.

Continuous wave lasers are most common for normal Raman spectroscopy, but pulsed lasers may also be used. These often have wider bandwidths than their CW counterparts but are very useful for other forms of Raman spectroscopy such as transient, time-resolved and resonance Raman.

Detectors

Raman scattered light is typically collected and either dispersed by a spectrograph or used with an interferometer for detection by Fourier Transform (FT) methods. In many cases commercially available FT-IR spectrometers can be modified to become FT-Raman spectrometers.

- **Detectors for Dispersive Raman:** In most cases, modern Raman spectrometers use array detectors such as CCDs. Various types of CCDs exist which are optimized for different wavelength ranges. Intensified CCDs can be used for very weak signals and/or pulsed lasers. The spectral range depends on the size of the CCD and the focal length of spectrograph used. It was once common to use monochromators coupled to photomultiplier tubes. In this case the monochromator would need to be moved in order to scan through a spectral range.
- **Detectors for FT-Raman:** FT-Raman is almost always used with NIR lasers and appropriate detectors must be used depending on the exciting wavelength. Germanium or Indium gallium arsenide (InGaAs) detectors are commonly used.

Filters

It is usually necessary to separate the Raman scattered light from the Rayleigh signal and reflected laser signal in order to collect high quality Raman spectra using a laser rejection filter. Notch or long-pass optical filters are typically used for this purpose. Before the advent of holographic filters it was common to use a triple-grating monochromator in subtractive mode to isolate the desired signal. This may still be used to record very small Raman shifts as holographic filters typically reflect some of the low frequency bands in addition to the unshifted laser light. However, Volume hologram filters are becoming more common which allow shifts as low as 5 cm^{-1} to be observed.

Applications

Raman spectroscopy is used in chemistry to identify molecules and study chemical bonding and intramolecular bonds. Because vibrational frequencies are specific to a molecule's chemical bonds and symmetry (the fingerprint region of organic molecules is in the wavenumber range 500–1,500 cm^{-1}), Raman provides a fingerprint to identify molecules. For instance, Raman and IR spectra were used to determine the vibrational frequencies of SiO, Si_2O_2, and Si_3O_3 on the basis of normal coordinate analyses. Raman is also used to study the addition of a substrate to an enzyme.

In solid-state physics, Raman spectroscopy is used to characterize materials, measure temperature, and find the crystallographic orientation of a sample. As with single molecules, a solid material can be identified by characteristic phonon modes. Information on the population of a phonon mode is given by the ratio of the Stokes and anti-Stokes intensity of the spontaneous Raman signal. Raman spectroscopy can also be used to observe other low frequency excitations of a solid, such as plasmons, magnons, and superconducting gap excitations. Distributed temperature sensing (DTS) uses the Raman-shifted backscatter from laser pulses to determine the temperature along optical fibers. The orientation of an anisotropic crystal can be found from the polarization of Raman-scattered light with respect to the crystal and the polarization of the laser light, if the crystal structure's point group is known.

In nanotechnology, a Raman microscope can be used to analyze nanowires to better understand their structures, and the radial breathing mode of carbon nanotubes is commonly used to evaluate their diameter.

Raman active fibers, such as aramid and carbon, have vibrational modes that show a shift in Raman frequency with applied stress. Polypropylene fibers exhibit similar shifts.

In solid state chemistry and the bio-pharmaceutical industry, Raman spectroscopy can be used to not only identify active pharmaceutical ingredients (APIs), but to identify their polymorphic forms, if more than one exist. For example, the drug Cayston (aztreonam), marketed by Gilead Sciences for cystic fibrosis, can be identified and characterized by IR and Raman spectroscopy. Using the correct polymorphic form in bio-pharmaceutical formulations is critical, since different forms have different physical properties, like solubility and melting point.

Raman spectroscopy has a wide variety of applications in biology and medicine. It has helped confirm the existence of low-frequency phonons in proteins and DNA, promoting studies of low-frequency collective motion in proteins and DNA and their biological functions. Raman reporter molecules with olefin or alkyne moieties are being developed for tissue imaging with SERS-labeled antibodies. Raman spectroscopy has also been used as a noninvasive technique for real-time, in situ biochemical characterization of wounds. Multivariate analysis of Raman spectra has enabled development of a quantitative measure for wound healing progress. Spatially offset Raman spectroscopy (SORS), which is less sensitive to surface layers than conventional Raman, can be used to discover counterfeit drugs without opening their packaging, and to non-invasively study biological tissue. A reason why Raman spectroscopy is useful in biological applications is because its results often do not face interference from water molecules, due to the fact that they have permanent dipole moments, and as a result, the Raman scattering cannot be picked up on. This is a large advantage, specifically in biological applications. Raman spectroscopy also has a wide usage for studying biominerals. Lastly, Raman gas analyzers have many practical applications, including real-time monitoring of anesthetic and respiratory gas mixtures during surgery.

Raman spectroscopy has been used in several research projects as a means to detect explosives from a safe distance using laser beams. Raman Spectroscopy is being further developed so it could be used in the clinical setting. Raman4Clinic is a European organization that is working on incorporating Raman Spectroscopy techniques in the medical field. They are currently working on different projects, one of them being monitoring cancer using bodily fluids such as urine and blood samples which are easily accessible. This technique would be less stressful on the patients than constantly having to take biopsies which are not always risk free.

In photovoltaics, Raman spectroscopy has gained more interest in the past few years demonstrating high efficacy in delivering important properties for such materials. This includes optoelectronic and physicochemical properties such as open circuit voltage, efficiency, and crystalline structure. This has been demonstrated with several photovoltaic technologies, including kesterite-based, CIGS devices, Monocrystalline silicon cells, and perovskites devices.

Art and Cultural Heritage

Raman spectroscopy is an efficient and non-destructive way to investigate works of art and cultural heritage artifacts, in part because it is a non-invasive process which can be applied *in situ*. It can be used to analyze the corrosion products on the surfaces of artifacts (statues, pottery, etc.), which can lend insight into the corrosive environments experienced by the artifacts. The resulting spectra can also be compared to the spectra of surfaces that are cleaned or intentionally corroded, which can aid in determining the authenticity of valuable historical artifacts.

It is capable of identifying individual pigments in paintings and their degradation products, which can provide insight into the working method of an artist in addition to aiding in authentication of paintings. It also gives information about the original state of the painting in cases where the pigments have degraded with age. Beyond the identification of pigments, extensive Raman microspectroscopic imaging has been shown to provide access to a plethora of trace compounds in Early Medieval Egyptian blue, which enable to reconstruct the individual "biography" of a colourant, including information on the type and provenance of the raw materials, synthesis and application of the pigment, and the ageing of the paint layer.

In addition to paintings and artifacts, Raman spectroscopy can be used to investigate the chemical composition of historical documents (such as the Book of Kells), which can provide insight about the social and economic conditions when they were created. It also offers a noninvasive way to determine the best method of preservation or conservation of such cultural heritage artifacts, by providing insight into the causes behind deterioration.

The IRUG (Infrared and Raman Users Group) Spectral Database is a rigorously peer-reviewed online database of IR and Raman reference spectra for cultural heritage materials such as works of art, architecture, and archaeological artifacts. The database is open for the general public to peruse, and includes interactive spectra for over a hundred different types of pigments and paints.

Microspectroscopy

Raman spectroscopy offers several advantages for microscopic analysis. Since it is a light scattering technique, specimens do not need to be fixed or sectioned. Raman spectra can be collected from a very small volume (< 1 μm in diameter, < 10 μm in depth); these spectra allow the identification of species present in that volume. Water does not generally interfere with Raman spectral analysis. Thus, Raman spectroscopy is suitable for the microscopic examination of minerals, materials such as polymers and ceramics, cells, proteins and forensic trace evidence.

A Raman microscope begins with a standard optical microscope, and adds an excitation laser, a monochromator or polychromator, and a sensitive detector (such as a charge-coupled device (CCD), or photomultiplier tube (PMT)). FT-Raman has also been used with microscopes, typically in combination with near-infrared (NIR) laser excitation. Ultraviolet microscopes and UV enhanced optics must be used when a UV laser source is used for Raman microspectroscopy.

In *direct imaging* (also termed *global imaging* or *wide-field illumination*), the whole field of view is examined for light scattering integrated over a small range of wavenumbers (Raman shifts).

For instance, a wavenumber characteristic for cholesterol could be used to record the distribution of cholesterol within a cell culture. This technique is being used for the characterization of large-scale devices, mapping of different compounds and dynamics study.

It has already been used for the characterization of graphene layers, J-aggregated dyes inside carbon nanotubes and multiple other 2D materials such as MoS_2 and WSe_2. Since the excitation beam is dispersed over the whole field of view, those measurements can be done without damaging the sample.

The most common approach is *hyperspectral imaging* or *chemical imaging*, in which thousands of Raman spectra are acquired from all over the field of view by, for example, raster scanning of a focused laser beam through a sample.

The data can be used to generate images showing the location and amount of different components. Having the full spectroscopic information available in every measurement spot has the advantage that several components can be mapped at the same time, including chemically similar and even polymorphic forms, which cannot be distinguished by detecting only one single wavenumber. Furthermore, material properties such as stress and strain, crystal orientation, crystallinity and incorporation of foreign ions into crystal lattices (e.g., doping, solid solution series) can be determined from hyperspectral maps. Taking the cell culture example, a hyperspectral image could show the distribution of cholesterol, as well as proteins, nucleic acids, and fatty acids. Sophisticated signal- and image-processing techniques can be used to ignore the presence of water, culture media, buffers, and other interferences.

Because a Raman microscope is a diffraction-limited system, its spatial resolution depends on the wavelength of light, the numerical aperture of the focusing element, and — in the case of confocal microscopy — on the diameter of the confocal aperture. When operated in the visible to near-infrared range, a Raman microscope can achieve lateral resolutions of approx. 1 µm down to 250 nm, depending on the wavelength and type of objective lens (e.g., air *vs.* water or oil immersion lenses). The depth resolution (if not limited by the optical penetration depth of the sample) can range from 1–6 µm with the smallest confocal pinhole aperture to tens of micrometers when operated without a confocal pinhole. Depending on the sample, the high laser power density due to microscopic focussing can have the benefit of enhanced photobleaching of molecules emitting interfering fluorescence. However, the laser wavelength and laser power have to be carefully selected for each type of sample to avoid its degradation.

Applications of Raman imaging range from materials sciences to biological studies. For each type of sample, the measurement parameters have to be individually optimized. For that reason, modern Raman microscopes are often equipped with several lasers offering different wavelengths, a set of objective lenses, and neutral density filters for tuning of the laser power reaching the sample. Selection of the laser wavelength mainly depends on optical properties of the sample and on the aim of the investigation. For example, Raman microscopy of biological and medical specimens is often performed using red to near-infrared excitation (e.g., 785 nm, or 1,064 nm wavelength). Due to typically low absorbances of biological samples in this spectral range, the risk of damaging the specimen as well as autofluorescence emission are reduced, and high penetration depths into tissues can be achieved. However, the intensity of Raman scattering at long wavelengths is low (owing to the ω^4 dependence of Raman scattering intensity), leading to long acquisition times. On the other hand, resonance Raman imaging of single-cell algae at 532 nm (green) can specifically probe the carotenoid distribution within a cell by a using low laser power of ~5 μW and only 100 ms acquisition time. Raman scattering, specifically tip-enhanced Raman spectroscopy, produces high resolution hyperspectral images of single molecules, atoms, and DNA.

Polarization Dependence of Raman Scattering

Raman scattering is polarization sensitive and can provide detailed information on symmetry of Raman active modes. While conventional Raman spectroscopy identifies chemical composition, polarization effects on Raman spectra can reveal information on the orientation of molecules in single crystals and anisotropic materials, e.g. strained plastic sheets, as well as the symmetry of vibrational modes.

Polarization–dependent Raman spectroscopy uses (plane) polarized laser excitation from a polarizer. The Raman scattered light collected is passed through a second polarizer (called the analyzer) before entering the detector. The analyzer is oriented either parallel or perpendicular to the polarization of the laser. Spectra acquired with the analyzer set at both perpendicular and parallel to the excitation plane can be used to calculate the depolarization ratio. Typically a polarization scrambler is placed between the analyzer and detector also.It is convenient in polarized Raman spectroscopy to describe the propagation and polarization directions using Porto's notation, described by and named after Brazilian physicist Sergio Pereira da Silva Porto.

For isotropic solutions, the Raman scattering from each mode either retains the polarization of the laser or becomes partly or fully depolarized. If the vibrational mode involved in the Raman scattering process is totally symmetric then the polarization of the Raman scattering will be the same as that of the incoming laser beam. In the case that the vibrational mode is not totally symmetric then the polarization will be lost (scrambled) partially or totally, which is referred to as depolarization. Hence polarized Raman spectroscopy can provide detailed information as to the symmetry labels of vibrational modes.

In the solid state, polarized Raman spectroscopy can be useful in the study of oriented samples such as single crystals. The polarizability of a vibrational mode is not equal along and across the bond. Therefore the intensity of the Raman scattering will be different when the laser's polarization is along and orthogonal to a particular bond axis. This effect can provide information on the orientation of molecules with a single crystal or material. The spectral information arising from this analysis is often used to understand macro-molecular orientation in crystal lattices, liquid crystals or polymer samples.

Characterization of the Symmetry of a Vibrational Mode

The polarization technique is useful in understanding the connections between molecular symmetry, Raman activity, and peaks in the corresponding Raman spectra. Polarized light in one direction only gives access to some Raman–active modes, but rotating the polarization gives access to other modes. Each mode is separated according to its symmetry.

The symmetry of a vibrational mode is deduced from the depolarization ratio ρ, which is the ratio of the Raman scattering with polarization orthogonal to the incident laser and the Raman scattering with the same polarization as the incident laser:

$$\rho = \frac{I_r}{I_u}$$

Raman Excitation Profile Analysis

Resonance Raman selection rules can be explained by the Kramers-Heisenberg-Dirac (KHD) equation using the Albrecht A and B terms, as demonstrated. The KHD expression is conveniently linked to the polarizability of the molecule within its frame of reference.

$$\left(\alpha_{\rho\sigma}\right)_{if} = \frac{1}{\hbar}\sum_{n}\left[\frac{\langle i|\mu_\rho|n\rangle\langle n|\mu_\sigma|f\rangle}{\omega_0+\omega_{nf}+i\Gamma_n}-\frac{\langle i|\mu_\sigma|n\rangle\langle n|\mu_\rho|f\rangle}{\omega_0-\omega_{ni}-i\Gamma_n}\right]$$

$$\equiv\langle i|\hat{\alpha}_{\rho\sigma}|f\rangle$$

The polarizability operator connecting the initial and final states expresses the transition polarizability as a matrix element, as a function of the incidence frequency ω_0. The directions x, y, and z in the molecular frame are represented by the Cartesian tensor ρ and σ here. Analyzing Raman excitation patterns requires the use of this equation, which is a sum-over-states expression for polarizability. This series of profiles illustrates the connection between a Raman active vibration's excitation frequency and intensity.

This method takes into account sums over Franck-Condon's active vibrational states and provides insight into electronic absorption and emission spectra. Nevertheless, the work highlights a flaw in the sum-over-states method, especially for large molecules like visible chromophores, which are commonly studied in Raman spectroscopy. The difficulty arises from the potentially infinite number of intermediary steps needed. While lowering the sum at higher vibrational states can help tiny molecules get over this issue, larger molecules find it more challenging when there are more terms in the sum, particularly in the condensed phase when individual eigenstates cannot be resolved spectrally.

To overcome this, two substitute techniques that do not require adding eigenstates can be considered. Among these two methods are available: the transform method. and Heller's time-dependent approach. The goal of both approaches is to take into consideration the frequency-dependent Raman cross-section $\sigma_R(\omega_0)$ of a particular normal mode.

Variants

At least 25 variations of Raman spectroscopy have been developed. The usual purpose is to enhance the sensitivity (e.g., surface-enhanced Raman), to improve the spatial resolution (Raman microscopy), or to acquire very specific information (resonance Raman).

Spontaneous (or Far-Field) Raman Spectroscopy

Terms such as *spontaneous Raman spectroscopy* or *normal Raman spectroscopy* summarize Raman spectroscopy techniques based on Raman scattering by using normal far-field optics as described above. Variants of normal Raman spectroscopy exist with respect to excitation-detection geometries, combination with other techniques, use of special (polarizing) optics and specific choice of excitation wavelengths for resonance enhancement.

- **Correlative Raman Imaging:** Raman microscopy can be combined with complementary imaging methods, such as atomic force microscopy (Raman-AFM) and scanning electron microscopy (Raman-SEM) to compare Raman distribution maps with (or overlay them onto) topographical or morphological images, and to correlate Raman spectra with complementary physical or chemical information (e.g., gained by SEM-EDX).
- **Resonance Raman Spectroscopy:** The excitation wavelength is matched to an electronic transition of the molecule or crystal, so that vibrational modes associated with the excited electronic state are greatly enhanced. This is useful for studying large molecules such as polypeptides, which might show hundreds of bands in "conventional" Raman spectra. It is also useful for associating normal modes with their observed frequency shifts.
- **Angle-Resolved Raman Spectroscopy:** Not only are standard Raman results recorded but also the angle with respect to the incident laser. If the orientation of the sample is known then detailed information about the phonon dispersion relation can also be gleaned from a single test.

- **Optical Tweezers Raman Spectroscopy (OTRS):** Used to study individual particles, and even biochemical processes in single cells trapped by optical tweezers.
- **Spatially Offset Raman Spectroscopy (SORS):** The Raman scattering beneath an obscuring surface is retrieved from a scaled subtraction of two spectra taken at two spatially offset points.
- **Raman Optical Activity (ROA):** Measures vibrational optical activity by means of a small difference in the intensity of Raman scattering from chiral molecules in right- and left-circularly polarized incident light or, equivalently, a small circularly polarized component in the scattered light.
- **Transmission Raman:** Allows probing of a significant bulk of a turbid material, such as powders, capsules, living tissue, etc. It was largely ignored following investigations in the late 1960s (Schrader and Bergmann, 1967) but was rediscovered in 2006 as a means of rapid assay of pharmaceutical dosage forms. There are medical diagnostic applications particularly in the detection of cancer.
- **Micro-Cavity Substrates:** A method that improves the detection limit of conventional Raman spectra using micro-Raman in a micro-cavity coated with reflective Au or Ag. The micro-cavity has a radius of several micrometers and enhances the entire Raman signal by providing multiple excitations of the sample and couples the forward-scattered Raman photons toward the collection optics in the back-scattered Raman geometry.
- **Stand-Off Remote Raman:** In standoff Raman, the sample is measured at a distance from the Raman spectrometer, usually by using a telescope for light collection. Remote Raman spectroscopy was proposed in the 1960s and initially developed for the measurement of atmospheric gases. The technique was extended In 1992 by Angel et al. for standoff Raman detection of hazardous inorganic and organic compounds.
- **X-Ray Raman Scattering:** Measures electronic transitions rather than vibrations.

Enhanced (or Near-Field) Raman Spectroscopy

Enhancement of Raman scattering is achieved by local electric-field enhancement by optical near-field effects (e.g. localized surface plasmons).

- **Surface-Enhanced Raman Spectroscopy (SERS):** Normally done in a silver or gold colloid or a substrate containing silver or gold. Surface plasmons of silver and gold are excited by the laser, resulting in an increase in the electric fields surrounding the metal. Given that Raman intensities are proportional to the electric field, there is large increase in the measured signal (by up to 10^{11}). This effect was originally observed by Martin Fleischmann but the prevailing explanation was proposed by Van Duyne in 1977. A comprehensive theory of the effect was given by Lombardi and Birke.

- **Surface-Enhanced Resonance Raman Spectroscopy (SERRS):** A combination of SERS and resonance Raman spectroscopy that uses proximity to a surface to increase Raman intensity, and excitation wavelength matched to the maximum absorbance of the molecule being analysed.
- **Tip-Enhanced Raman Spectroscopy (TERS):** TERS combines the chemical sensitivity of SERS with the high spatial resolution of scanning probe microscopy techniques, enabling chemical imaging of surfaces at the nanometre length-scale with high detection sensitivity. It uses a metallic (usually silver-/gold-coated AFM or STM) tip to enhance the Raman signals of molecules situated in its vicinity. The spatial resolution is approximately the size of the tip apex (20–30 nm). TERS has been shown to have sensitivity down to the single molecule level and holds some promise for bioanalysis applications and DNA sequencing. TERS was used to image the vibrational normal modes of single molecules.
- **Surface Plasmon Polariton Enhanced Raman Scattering (SPPERS):** This approach exploits apertureless metallic conical tips for near field excitation of molecules. This technique differs from the TERS approach due to its inherent capability of suppressing the background field. In fact, when an appropriate laser source impinges on the base of the cone, a TM0 mode (polaritonic mode) can be locally created, namely far away from the excitation spot (apex of the tip). The mode can propagate along the tip without producing any radiation field up to the tip apex where it interacts with the molecule. In this way, the focal plane is separated from the excitation plane by a distance given by the tip length, and no background plays any role in the Raman excitation of the molecule.

Non-Linear Raman Spectroscopy

Raman signal enhancements are achieved through non-linear optical effects, typically realized by mixing two or more wavelengths emitted by spatially and temporally synchronized pulsed lasers.

- **Hyper Raman:** A non-linear effect in which the vibrational modes interact with the second harmonic of the excitation beam. This requires very high power, but allows the observation of vibrational modes that are normally "silent". It frequently relies on SERS-type enhancement to boost the sensitivity.

- **Stimulated Raman Spectroscopy (SRS):** A pump-probe technique, where a spatially coincident, two color pulse (with polarization either parallel or perpendicular) transfers the population from ground to a rovibrationally excited state. If the difference in energy corresponds to an allowed Raman transition, scattered light will correspond to loss or gain in the pump beam.
- **Inverse Raman Spectroscopy:** A synonym for stimulated Raman loss spectroscopy.
- **Coherent Anti-Stokes Raman Spectroscopy (CARS):** Two laser beams are used to generate a coherent anti-Stokes frequency beam, which can be enhanced by resonance.

Morphologically-Directed Raman Spectroscopy

Morphologically Directed Raman Spectroscopy (MDRS) combines automated particle imaging and Raman microspectroscopy into a singular integrated platform in order to provide particle size, shape, and chemical identification. Automated particle imaging determines the particle size and shape distributions of components within a blended sample from images of individual particles. The information gathered from automated particle imaging is then utilized to direct the Raman spectroscopic analysis. The Raman spectroscopic analytical process is performed on a randomly-selected subset of the particles, allowing chemical identification of the sample's multiple components. Tens of thousands of particles can be imaged in a matter of minutes using the MDRS method, making the process ideal for forensic analysis and investigating counterfeit pharmaceuticals and subsequent adjudications.

5.3.3 Nuclear Magnetic Resonance (NMR) Spectroscopy

Nuclear magnetic resonance spectroscopy, most commonly known as NMR spectroscopy or magnetic resonance spectroscopy (MRS), is a spectroscopic technique based on re-orientation of atomic nuclei with non-zero nuclear spins in an external magnetic field. This re-orientation occurs with absorption of electromagnetic radiation in the radio frequency region from roughly 4 to 900 MHz, which depends on the isotopic nature of the nucleus and increased proportionally to the strength of the external magnetic field. Notably, the resonance frequency of each NMR-active nucleus depends on its chemical environment. As a result, NMR spectra provide information about individual functional groups present in the sample, as well as about connections between nearby nuclei in the same molecule. As the NMR spectra are unique or highly characteristic to individual compounds and functional groups, NMR spectroscopy is one of the most important methods to identify molecular structures, particularly of organic compounds.

The principle of NMR usually involves three sequential steps:

- The alignment (polarization) of the magnetic nuclear spins in an applied, constant magnetic field B_0.
- The perturbation of this alignment of the nuclear spins by a weak oscillating magnetic field, usually referred to as a radio-frequency (RF) pulse.
- Detection and analysis of the electromagnetic waves emitted by the nuclei of the sample as a result of this perturbation.

Similarly, biochemists use NMR to identify proteins and other complex molecules. Besides identification, NMR spectroscopy provides detailed information about the structure, dynamics, reaction state, and chemical environment of molecules. The most common types of NMR are proton and carbon-13 NMR spectroscopy, but it is applicable to any kind of sample that contains nuclei possessing spin.

NMR spectra are unique, well-resolved, analytically tractable and often highly predictable for small molecules. Different functional groups are obviously distinguishable, and identical functional groups with differing neighboring substituents still give distinguishable signals. NMR has largely replaced traditional wet chemistry tests such as color reagents or typical chromatography for identification.

The most significant drawback of NMR spectroscopy is its poor sensitivity (compared to other analytical methods, such as mass spectrometry). Typically 2–50 mg of a substance is required to record a decent-quality NMR spectrum. The NMR method is non-destructive, thus the substance may be recovered. To obtain high-resolution NMR spectra, solid substances are usually dissolved to make liquid solutions, although solid-state NMR spectroscopy is also possible.

The timescale of NMR is relatively long, and thus it is not suitable for observing fast phenomena, producing only an averaged spectrum. Although large amounts of impurities do show on an NMR spectrum, better methods exist for detecting impurities, as NMR is inherently not very sensitive – though at higher frequencies, sensitivity is higher.

Correlation spectroscopy is a development of ordinary NMR. In two-dimensional NMR, the emission is centered around a single frequency, and correlated resonances are observed.

This allows identifying the neighboring substituents of the observed functional group, allowing unambiguous identification of the resonances. There are also more complex 3D and 4D methods and a variety of methods designed to suppress or amplify particular types of resonances. In nuclear Overhauser effect (NOE) spectroscopy, the relaxation of the resonances is observed. As NOE depends on the proximity of the nuclei, quantifying the NOE for each nucleus allows construction of a three-dimensional model of the molecule.

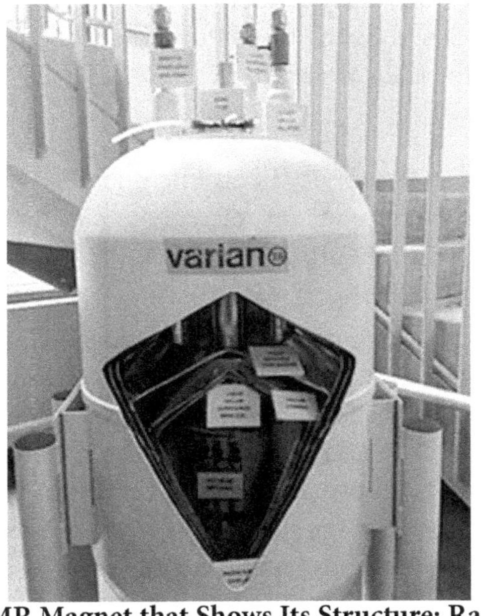

Fig 1: Cutaway of an NMR Magnet that Shows Its Structure: Radiation Shield, Vacuum Chamber, Liquid-Nitrogen Vessel, Liquid-Helium Vessel, and Cryogenic Shims

NMR spectrometers are relatively expensive; universities usually have them, but they are less common in private companies. Between 2000 and 2015, an NMR spectrometer cost around 0.5–5 million USD.

Modern NMR spectrometers have a very strong, large and expensive liquid-helium-cooled superconducting magnet, because resolution directly depends on magnetic field strength. Higher magnetic field also improves the sensitivity of the NMR spectroscopy, which depends on the population difference between the two nuclear levels, which increases exponentially with the magnetic field strength.

Less expensive machines using permanent magnets and lower resolution are also available, which still give sufficient performance for certain applications such as reaction monitoring and quick checking of samples.

There are even benchtop nuclear magnetic resonance spectrometers. NMR spectra of protons (^{1}H nuclei) can be observed even in Earth magnetic field. Low-resolution NMR produces broader peaks, which can easily overlap one another, causing issues in resolving complex structures. The use of higher-strength magnetic fields result in a better sensitivity and higher resolution of the peaks, and it is preferred for research purposes.

History

Credit for the discovery of NMR goes to Isidor Isaac Rabi, who received the Nobel Prize in Physics in 1944. The Purcell group at Harvard University and the Bloch group at Stanford University independently developed NMR spectroscopy in the late 1940s and early 1950s. Edward Mills Purcell and Felix Bloch shared the 1952 Nobel Prize in Physics for their inventions.

NMR-Active Criteria

The key determinant of NMR activity in atomic nuclei is the nuclear spin quantum number (I). This intrinsic quantum property, similar to an atom's "spin", characterizes the angular momentum of the nucleus. To be NMR-active, a nucleus must have a non-zero nuclear spin ($I \neq 0$).

It is this non-zero spin that enables nuclei to interact with external magnetic fields and show signals in NMR. Atoms with an odd sum of protons and neutrons exhibit half-integer values for the nuclear spin quantum number (I = 1/2, 3/2, 5/2, and so on). These atoms are NMR-active because they possess non-zero nuclear spin. Atoms with an even sum but both an odd number of protons and an odd number of neutrons exhibit integer nuclear spins (I = 1, 2, 3, and so on). Conversely, atoms with an even number of both protons and neutrons have a nuclear spin quantum number of zero (I = 0), and therefore are not NMR-active. NMR-active nuclei, particularly those with a spin quantum number of 1/2, are of great significance in NMR spectroscopy. Examples include ^{1}H, ^{13}C, ^{15}N, and ^{31}P.

Main Aspects of NMR Techniques

Resonant Frequency

When placed in a magnetic field, NMR active nuclei (such as ^{1}H or ^{13}C) absorb electromagnetic radiation at a frequency characteristic of the isotope. The resonant frequency, energy of the radiation absorbed, and the intensity of the signal are proportional to the strength of the magnetic field. For example, in a 21-tesla magnetic field, hydrogen nuclei (protons) resonate at 900 MHz. It is common to refer to a 21 T magnet as a 900 MHz magnet, since hydrogen is the most common nucleus detected. However, different nuclei will resonate at different frequencies at this field strength in proportion to their nuclear magnetic moments.

Sample Handling

An NMR spectrometer typically consists of a spinning sample-holder inside a very strong magnet, a radio-frequency emitter, and a receiver with a probe (an antenna assembly) that goes inside the magnet to surround the sample, optionally gradient coils for diffusion measurements, and electronics to control the system. Spinning the sample is usually necessary to average out diffusional motion, however, some experiments call for a stationary sample when solution movement is an important variable. For instance, measurements of diffusion constants (*diffusion ordered spectroscopy* or DOSY) are done using a stationary sample with spinning off, and flow cells can be used for online analysis of process flows.

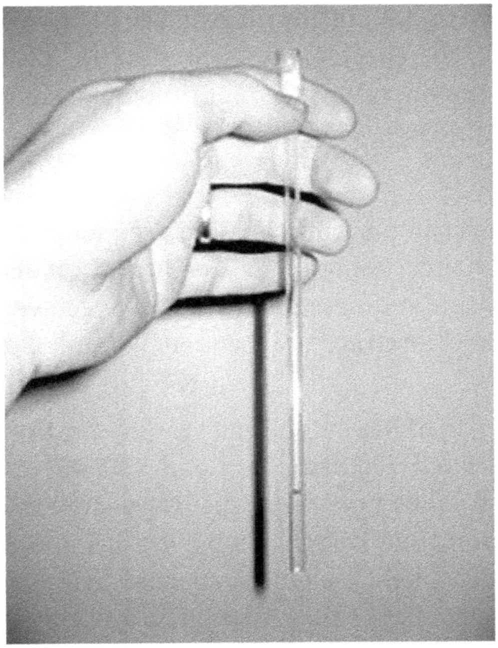

Fig 2: The NMR Sample is Prepared in a Thin-Walled Glass Tube—An NMR Tube

Deuterated Solvents

The vast majority of molecules in a solution are solvent molecules, and most regular solvents are hydrocarbons and so contain NMR-active hydrogen-1 nuclei. In order to avoid having the signals from solvent hydrogen atoms overwhelm the experiment and interfere in analysis of the dissolved analyte, deuterated solvents are used where >99% of the protons are replaced with deuterium (hydrogen-2). The most widely used deuterated solvent is deuterochloroform ($CDCl_3$), although other solvents may be used for various reasons, such as solubility of a sample, desire to control hydrogen bonding, or melting or boiling points. The chemical shifts of a molecule change slightly between solvents, and therefore the solvent used is almost always reported with chemical shifts. Proton NMR spectra are often calibrated against the known solvent residual proton peak as an internal standard instead of adding tetramethylsilane (TMS), which is conventionally defined as having a chemical shift of zero.

Shim and Lock

To detect the very small frequency shifts due to nuclear magnetic resonance, the applied magnetic field must be extremely uniform throughout the sample volume. High-resolution NMR spectrometers use shims to adjust the homogeneity of the magnetic field to parts per billion (ppb) in a volume of a few cubic centimeters. In order to detect and compensate for inhomogeneity and drift in the magnetic field, the spectrometer maintains a "lock" on the solvent deuterium frequency with a separate lock unit, which is essentially an additional transmitter and RF processor tuned to the lock nucleus (deuterium) rather than the nuclei of the sample of interest. In modern NMR spectrometers shimming is adjusted automatically, though in some cases the operator has to optimize the shim parameters manually to obtain the best possible resolution.

Acquisition of Spectra

Upon excitation of the sample with a radio frequency (60–1000 MHz) pulse, a nuclear magnetic resonance response – a free induction decay (FID) – is obtained. It is a very weak signal and requires sensitive radio receivers to pick up. A Fourier transform is carried out to extract the frequency-domain spectrum from the raw time-domain FID.

A spectrum from a single FID has a low signal-to-noise ratio, but it improves readily with averaging of repeated acquisitions. Good ^{1}H NMR spectra can be acquired with 16 repeats, which takes only minutes. However, for elements heavier than hydrogen, the relaxation time is rather long, e.g. around 8 seconds for ^{13}C. Thus, acquisition of quantitative heavy-element spectra can be time-consuming, taking tens of minutes to hours.

Following the pulse, the nuclei are, on average, excited to a certain angle vs. the spectrometer magnetic field. The extent of excitation can be controlled with the pulse width, typically about 3–8 μs for the optimal 90° pulse. The pulse width can be determined by plotting the (signed) intensity as a function of pulse width. It follows a sine curve and, accordingly, changes sign at pulse widths corresponding to 180° and 360° pulses.

Decay times of the excitation, typically measured in seconds, depend on the effectiveness of relaxation, which is faster for lighter nuclei and in solids, slower for heavier nuclei and in solutions, and can be very long in gases. If the second excitation pulse is sent prematurely before the relaxation is complete, the average magnetization vector has not decayed to ground state, which affects the strength of the signal in an unpredictable manner.

In practice, the peak areas are then not proportional to the stoichiometry; only the presence, but not the amount of functional groups is possible to discern. An inversion recovery experiment can be done to determine the relaxation time and thus the required delay between pulses. A 180° pulse, an adjustable delay, and a 90° pulse is transmitted. When the 90° pulse exactly cancels out the signal, the delay corresponds to the time needed for 90° of relaxation. Inversion recovery is worthwhile for quantitative ^{13}C, ^{2}D and other time-consuming experiments.

Spectral Interpretation

NMR signals are ordinarily characterized by three variables: chemical shift, spin-spin coupling, and relaxation time.

Chemical Shift

The energy difference ΔE between nuclear spin states is proportional to the magnetic field (Zeeman effect). ΔE is also sensitive to electronic environment of the nucleus, giving rise to what is known as the chemical shift, δ. The simplest types of NMR graphs are plots of the different chemical shifts of the nuclei being studied in the molecule. The value of δ is often expressed in terms of "shielding": shielded nuclei have higher ΔE. The range of δ values is called the dispersion. It is rather small for ^{1}H signals, but much larger for other nuclei. NMR signals are reported relative to a reference signal, usually that of TMS (tetramethylsilane). Additionally, since the distribution of NMR signals is field-dependent, these frequencies are divided by the spectrometer frequency. However, since we are dividing Hz by MHz, the resulting number would be too small, and thus it is multiplied by a million. This operation therefore gives a locator number called the "chemical shift" with units of parts per million. The chemical shift provides structural information.

The conversion of chemical shifts is called *assigning* the spectrum. For diamagnetic organic compounds, assignments of ^{1}H and ^{13}C NMR spectra are extremely sophisticated because of the large databases and easy computational tools. In general, chemical shifts for protons are highly predictable, since the shifts are primarily determined by shielding effects (electron density). The chemical shifts for many heavier nuclei are more strongly influenced by other factors, including excited states ("paramagnetic" contribution to shielding tensor). This paramagnetic contribution, which is unrelated to paramagnetism) not only disrupts trends in chemical shifts, which complicates assignments, but it also gives rise to very large chemical shift ranges. For example, most ^{1}H NMR signals for most organic compounds are within 15 ppm. For ^{31}P NMR, the range is hundreds of ppm.

In paramagnetic NMR spectroscopy, the samples are paramagnetic, i.e. they contain unpaired electrons. The paramagnetism gives rise to very diverse chemical shifts. In ^{1}H NMR spectroscopy, the chemical shift range can span up to thousands of ppm.

J-Coupling

Some of the most useful information for structure determination in a one-dimensional NMR spectrum comes from J-coupling, or scalar coupling (a special case of spin–spin coupling), between NMR active nuclei. This coupling arises from the interaction of different spin states through the chemical bonds of a molecule and results in the splitting of NMR signals. For a proton, the local magnetic field is slightly different depending on whether an adjacent nucleus points towards or against the spectrometer magnetic field, which gives rise to two signals per proton instead of one. These splitting patterns can be complex or simple and, likewise, can be straightforwardly interpretable or deceptive. This coupling provides detailed insight into the connectivity of atoms in a molecule.

The multiplicity of the splitting is an effect of the spins of the nuclei that are coupled and the number of such nuclei involved in the coupling. Coupling to *n* equivalent spin-1/2 nuclei splits the signal into a $n + 1$ multiplet with intensity ratios following Pascal's triangle as described in the table. Coupling to additional spins leads to further splittings of each component of the multiplet, e.g. coupling to two different spin-1/2 nuclei with significantly different coupling constants leads to a *doublet of doublets* (abbreviation: dd). Note that coupling between nuclei that are chemically equivalent (that is, have the same chemical shift) has no effect on the NMR spectra, and couplings between nuclei that are distant (usually more than 3 bonds apart for protons in flexible molecules) are usually too small to cause observable splittings. *Long-range* couplings over more than three bonds can often be observed in cyclic and aromatic compounds, leading to more complex splitting patterns.

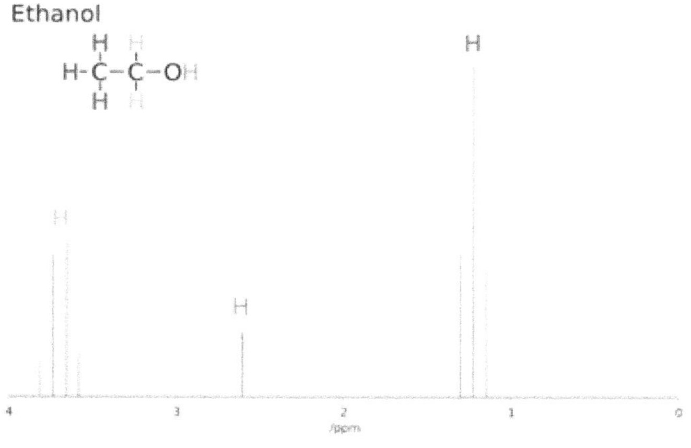

Fig 3: Example ^{1}H NMR Spectrum (1-Dimensional) of Ethanol Plotted as Signal Intensity vs. Chemical Shift.

For example, in the proton spectrum for ethanol, the CH_3 group is split into a *triplet* with an intensity ratio of 1:2:1 by the two neighboring CH_2 protons. Similarly, the

CH_2 is split into a *quartet* with an intensity ratio of 1:3:3:1 by the three neighboring CH_3 protons. In principle, the two CH_2 protons would also be split again into a *doublet* to form a *doublet of quartets* by the hydroxyl proton, but intermolecular exchange of the acidic hydroxyl proton often results in a loss of coupling information.

Coupling to any spin-1/2 nuclei such as phosphorus-31 or fluorine-19 works in this fashion (although the magnitudes of the coupling constants may be very different). But the splitting patterns differ from those described above for nuclei with spin greater than 1/2 because the spin quantum number has more than two possible values. For instance, coupling to deuterium (a spin-1 nucleus) splits the signal into a *1:1:1 triplet* because the spin 1 has three spin states. Similarly, a spin-3/2 nucleus such as ^{35}Cl splits a signal into a *1:1:1:1 quartet* and so on.

Coupling combined with the chemical shift (and the integration for protons) tells us not only about the chemical environment of the nuclei, but also the number of *neighboring* NMR active nuclei within the molecule.

In more complex spectra with multiple peaks at similar chemical shifts or in spectra of nuclei other than hydrogen, coupling is often the only way to distinguish different nuclei.

The magnitude of the coupling (the coupling constant *J*) is an effect of how strongly the nuclei are coupled to each other. For simple cases, this is an effect of the bonding distance between the nuclei, the magnetic moment of the nuclei, and the dihedral angle between them.

Second-Order (or Strong) Coupling

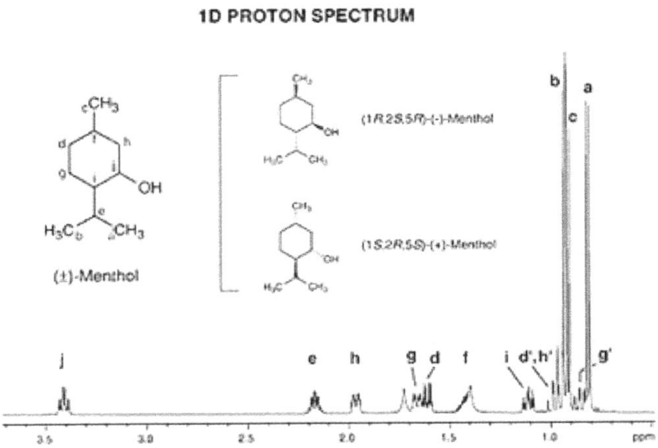

Fig 4: ¹H NMR Spectrum of Menthol with Chemical Shift in PPM on the Horizontal Axis

The above description assumes that the coupling constant is small in comparison with the difference in NMR frequencies between the inequivalent spins. If the shift separation decreases (or the coupling strength increases), the multiplet intensity patterns are first distorted, and then become more complex and less easily analyzed (especially if more than two spins are involved).

Intensification of some peaks in a multiplet is achieved at the expense of the remainder, which sometimes almost disappear in the background noise, although the integrated area under the peaks remains constant. In most high-field NMR, however, the distortions are usually modest, and the characteristic distortions (*roofing*) can in fact help to identify related peaks.

Some of these patterns can be analyzed with the method published by John Pople, though it has limited scope.

Second-order effects decrease as the frequency difference between multiplets increases, so that high-field (i.e. high-frequency) NMR spectra display less distortion than lower-frequency spectra. Early spectra at 60 MHz were more prone to distortion than spectra from later machines typically operating at frequencies at 200 MHz or above.

Furthermore, as in the figure to the right, J-coupling can be used to identify ortho-meta-para substitution of a ring. Ortho coupling is the strongest at 15 Hz, Meta follows with an average of 2 Hz, and finally para coupling is usually insignificant for studies.

Magnetic Inequivalence

More subtle effects can occur if chemically equivalent spins (i.e., nuclei related by symmetry and so having the same NMR frequency) have different coupling relationships to external spins. Spins that are chemically equivalent but are not indistinguishable (based on their coupling relationships) are termed magnetically inequivalent. For example, the 4 H sites of 1,2-dichlorobenzene divide into two chemically equivalent pairs by symmetry, but an individual member of one of the pairs has different couplings to the spins making up the other pair.

Magnetic inequivalence can lead to highly complex spectra, which can only be analyzed by computational modeling. Such effects are more common in NMR spectra of aromatic and other non-flexible systems, while conformational averaging about C–C bonds in flexible molecules tends to equalize the couplings between protons on adjacent carbons, reducing problems with magnetic inequivalence.

Correlation Spectroscopy

Correlation spectroscopy is one of several types of two-dimensional nuclear magnetic resonance (NMR) spectroscopy or 2D-NMR. This type of NMR experiment is best known by its acronym, COSY. Other types of two-dimensional NMR include J-spectroscopy, exchange spectroscopy (EXSY), Nuclear Overhauser effect spectroscopy (NOESY), total correlation spectroscopy (TOCSY), and heteronuclear correlation experiments, such as HSQC, HMQC, and HMBC. In correlation spectroscopy, emission is centered on the peak of an individual nucleus; if its magnetic field is correlated with another nucleus by through-bond (COSY, HSQC, etc.) or through-space (NOE) coupling, a response can also be detected on the frequency of the correlated nucleus. Two-dimensional NMR spectra provide more information about a molecule than one-dimensional NMR spectra and are especially useful in determining the structure of a molecule, particularly for molecules that are too complicated to work with using one-dimensional NMR. The first two-dimensional experiment, COSY, was proposed by Jean Jeener, a professor at Université Libre de Bruxelles, in 1971. This experiment was later implemented by Walter P. Aue, Enrico Bartholdi and Richard R. Ernst, who published their work in 1976.

Solid-State Nuclear Magnetic Resonance

A variety of physical circumstances do not allow molecules to be studied in solution, and at the same time not by other spectroscopic techniques to an atomic level, either. In solid-phase media, such as crystals, microcrystalline powders, gels, anisotropic solutions, etc., it is in particular the dipolar coupling and chemical shift anisotropy that become dominant to the behaviour of the nuclear spin systems. In conventional solution-state NMR spectroscopy, these additional interactions would lead to a significant broadening of spectral lines. A variety of techniques allows establishing high-resolution conditions, that can, at least for ^{13}C spectra, be comparable to solution-state NMR spectra.

Two important concepts for high-resolution solid-state NMR spectroscopy are the limitation of possible molecular orientation by sample orientation, and the reduction of anisotropic nuclear magnetic interactions by sample spinning. Of the latter approach, fast spinning around the magic angle is a very prominent method, when the system comprises spin-1/2 nuclei. Spinning rates of about 20 kHz are used, which demands special equipment. A number of intermediate techniques, with samples of partial alignment or reduced mobility, is currently being used in NMR spectroscopy.

Applications in which solid-state NMR effects occur are often related to structure investigations on membrane proteins, protein fibrils or all kinds of polymers, and chemical analysis in inorganic chemistry, but also include "exotic" applications like the plant leaves and fuel cells. For example, Rahmani et al. studied the effect of pressure and temperature on the bicellar structures' self-assembly using deuterium NMR spectroscopy.

Biomolecular NMR Spectroscopy

Proteins

Much of the innovation within NMR spectroscopy has been within the field of protein NMR spectroscopy, an important technique in structural biology. A common goal of these investigations is to obtain high resolution 3-dimensional structures of the protein, similar to what can be achieved by X-ray crystallography. In contrast to X-ray crystallography, NMR spectroscopy is usually limited to proteins smaller than 35 kDa, although larger structures have been solved. NMR spectroscopy is often the only way to obtain high resolution information on partially or wholly intrinsically unstructured proteins. It is now a common tool for the determination of Conformation Activity Relationships where the structure before and after interaction with, for example, a drug candidate is compared to its known biochemical activity.

Because of the much higher number of atoms present in a protein molecule in comparison with a small organic compound, the basic 1D spectra become crowded with overlapping signals to an extent where direct spectral analysis becomes untenable. Therefore, multidimensional (2, 3 or 4D) experiments have been devised to deal with this problem. To facilitate these experiments, it is desirable to isotopically label the protein with ^{13}C and ^{15}N because the predominant naturally occurring isotope ^{12}C is not NMR-active and the nuclear quadrupole moment of the predominant naturally occurring ^{14}N isotope prevents high resolution information from being obtained from this nitrogen isotope.

The most important method used for structure determination of proteins utilizes NOE experiments to measure distances between atoms within the molecule. Subsequently, the distances obtained are used to generate a 3D structure of the molecule by solving a distance geometry problem. NMR can also be used to obtain information on the dynamics and conformational flexibility of different regions of a protein.

Nucleic Acids

Nucleic acid NMR is the use of NMR spectroscopy to obtain information about the structure and dynamics of polynucleic acids, such as DNA or RNA. As of 2003, nearly half of all known RNA structures had been determined by NMR spectroscopy.

Nucleic acid and protein NMR spectroscopy are similar but differences exist. Nucleic acids have a smaller percentage of hydrogen atoms, which are the atoms usually observed in NMR spectroscopy, and because nucleic acid double helices are stiff and roughly linear, they do not fold back on themselves to give "long-range" correlations. The types of NMR usually done with nucleic acids are ^{1}H or proton NMR, ^{13}C NMR, ^{15}N NMR, and ^{31}P NMR. Two-dimensional NMR methods are almost always used, such as correlation spectroscopy (COSY) and total coherence transfer spectroscopy (TOCSY) to detect through-bond nuclear couplings, and nuclear Overhauser effect spectroscopy (NOESY) to detect couplings between nuclei that are close to each other in space.

Parameters taken from the spectrum, mainly NOESY cross-peaks and coupling constants, can be used to determine local structural features such as glycosidic bond angles, dihedral angles (using the Karplus equation), and sugar pucker conformations. For large-scale structure, these local parameters must be supplemented with other structural assumptions or models, because errors add up as the double helix is traversed, and unlike with proteins, the double helix does not have a compact interior and does not fold back upon itself.

NMR is also useful for investigating nonstandard geometries such as bent helices, non-Watson–Crick basepairing, and coaxial stacking. It has been especially useful in probing the structure of natural RNA oligonucleotides, which tend to adopt complex conformations such as stem-loops and pseudoknots. NMR is also useful for probing the binding of nucleic acid molecules to other molecules, such as proteins or drugs, by seeing which resonances are shifted upon binding of the other molecule.

Carbohydrates

Carbohydrate NMR spectroscopy addresses questions on the structure and conformation of carbohydrates. The analysis of carbohydrates by 1H NMR is challenging due to the limited variation in functional groups, which leads to 1H resonances concentrated in narrow bands of the NMR spectrum. In other words, there is poor spectral dispersion. The anomeric proton resonances are segregated from the others due to fact that the anomeric carbons bear two oxygen atoms. For smaller carbohydrates, the dispersion of the anomeric proton resonances facilitates the use of 1D TOCSY experiments to investigate the entire spin systems of individual carbohydrate residues.

Drug Discovery

Knowledge of energy minima and rotational energy barriers of small molecules in solution can be found using NMR, e.g. looking at free ligand conformational preferences and conformational dynamics, respectively. This can be used to guide drug design hypotheses, since experimental and calculated values are comparable. For example, AstraZeneca uses NMR for its oncology research & development.

High-Pressure NMR Spectroscopy

One of the first scientific works devoted to the use of pressure as a variable parameter in NMR experiments was the work of J. Jonas published in the journal Annual Review of Biophysics in 1994. The use of high pressures in NMR spectroscopy was primarily driven by the desire to study biochemical systems, where the use of high pressure allows controlled changes in intermolecular interactions without significant perturbations.

Of course, attempts have been made to solve scientific problems using high-pressure NMR spectroscopy. However, most of them were difficult to reproduce due to the problem of equipment for creating and maintaining high pressure. In the most common types of NMR cells for realization of high-pressure NMR experiments are given.

High-pressure NMR spectroscopy has been widely used for a variety of applications, mainly related to the characterization of the structure of protein molecules.

However, in recent years, software and design solutions have been proposed to characterize the chemical and spatial structures of small molecules in a supercritical fluid environment, using state parameters as a driving force for such changes.

5.3.4 X-ray Fluorescence (XRF) Spectroscopy

X-ray Fluorescence (XRF) spectroscopy is an advanced technique that analyzes the elemental composition of a sample by measuring the characteristic X-rays emitted when the sample is exposed to high-energy X-rays. XRF spectroscopy is widely used in forensic chemistry for the analysis of trace elements in a variety of materials, including metals, glass, soil, and paint.

XRF spectroscopy is highly sensitive and can detect elements at very low concentrations, making it ideal for analyzing trace evidence. For example, XRF can be used to examine gunshot residue (GSR) on a suspect's hands or clothing, helping to determine whether they recently fired a weapon. It can also be used to analyze the elemental composition of paint samples, linking them to specific vehicles or buildings.

One of the key advantages of XRF spectroscopy is its non-destructive nature, allowing forensic scientists to preserve the integrity of the evidence for further analysis. It is also a rapid technique, providing results within minutes, which is particularly valuable in time-sensitive investigations.

5.3.5 Inductively Coupled Plasma Mass Spectrometry (ICP-MS)

Inductively coupled plasma mass spectrometry (ICP-MS) is a type of mass spectrometry that uses an inductively coupled plasma to ionize the sample. It atomizes the sample and creates atomic and small polyatomic ions, which are then detected.

It is known and used for its ability to detect metals and several non-metals in liquid samples at very low concentrations. It can detect different isotopes of the same element, which makes it a versatile tool in isotopic labeling.

Compared to atomic absorption spectroscopy, ICP-MS has greater speed, precision, and sensitivity. However, compared with other types of mass spectrometry, such as thermal ionization mass spectrometry (TIMS) and glow discharge mass spectrometry (GD-MS), ICP-MS introduces many interfering species: argon from the plasma, component gases of air that leak through the cone orifices, and contamination from glassware and the cones.

Components

Inductively Coupled Plasma

An inductively coupled plasma is a plasma that is energized (ionized) by inductively heating the gas with an electromagnetic coil, and contains a sufficient concentration of ions and electrons to make the gas electrically conductive. Not all of the gas needs to be ionized for the gas to have the characteristics of a plasma; as little as 1% ionization creates a plasma. The plasmas used in spectrochemical analysis are essentially electrically neutral, with each positive charge on an ion balanced by a free electron. In these plasmas the positive ions are almost all singly charged and there are few negative ions, so there are nearly equal numbers of ions and electrons in each unit volume of plasma.

The ICPs have two operation modes, called capacitive (E) mode with low plasma density and inductive (H) mode with high plasma density, and E to H heating mode transition occurs with external inputs. The Inductively Coupled Plasma Mass Spectrometry is operated in the H mode.

What makes Inductively Coupled Plasma Mass Spectrometry (ICP-MS) unique to other forms of inorganic mass spectrometry is its ability to sample the analyte continuously, without interruption.

This is in contrast to other forms of inorganic mass spectrometry; Glow Discharge Mass Spectrometry (GDMS) and Thermal Ionization Mass Spectrometry (TIMS), that require a two-stage process: Insert sample(s) into a vacuum chamber, seal the vacuum chamber, pump down the vacuum, energize sample, thereby sending ions into the mass analyzer. With ICP-MS the sample to be analyzed is sitting at atmospheric pressure.

Through the effective use of differential pumping; multiple vacuum stages separate by differential apertures (holes), the ions created in the argon plasma are, with the aid of various electrostatic focusing techniques, transmitted through the mass analyzer to the detector(s) and counted. Not only does this enable the analyst to radically increase sample throughput (amount of samples over time), but has also made it possible to do what is called "time resolved acquisition".

Hyphenated techniques like Liquid Chromatography ICP-MS (LC-ICP-MS); Laser Ablation ICP-MS (LA-ICP-MS); Flow Injection ICP-MS (FIA-ICP-MS), etc. have benefited from this relatively new technology. It has stimulated the development new tools for research including geochemistry and forensic chemistry; biochemistry and oceanography.

Additionally, increases in sample throughput from dozens of samples a day to hundreds of samples a day have revolutionized environmental analysis, reducing costs. Fundamentally, this is all due to the fact that while the sample resides at environmental pressure, the analyzer and detector are at 1/10,000,000 of that same pressure during normal operation.

An inductively coupled plasma (ICP) for spectrometry is sustained in a torch that consists of three concentric tubes, usually made of quartz, although the inner tube (injector) can be sapphire if hydrofluoric acid is being used. The end of this torch is placed inside an induction coil supplied with a radio-frequency electric current. A flow of argon gas (usually 13 to 18 liters per minute) is introduced between the two outermost tubes of the torch and an electric spark is applied for a short time to introduce free electrons into the gas stream.

These electrons interact with the radio-frequency magnetic field of the induction coil and are accelerated first in one direction, then the other, as the field changes at high frequency (usually 27.12 million cycles per second). The accelerated electrons collide with argon atoms, and sometimes a collision causes an argon atom to part with one of its electrons. The released electron is in turn accelerated by the rapidly changing magnetic field. The process continues until the rate of release of new electrons in collisions is balanced by the rate of recombination of electrons with argon ions (atoms that have lost an electron).

This produces a 'fireball' that consists mostly of argon atoms with a rather small fraction of free electrons and argon ions. The temperature of the plasma is very high, of the order of 10,000 K. The plasma also produces ultraviolet light, so for safety should not be viewed directly.

The ICP can be retained in the quartz torch because the flow of gas between the two outermost tubes keeps the plasma away from the walls of the torch. A second flow of argon (around 1 liter per minute) is usually introduced between the central tube and the intermediate tube to keep the plasma away from the end of the central tube. A third flow (again usually around 1 liter per minute) of gas is introduced into the central tube of the torch. This gas flow passes through the centre of the plasma, where it forms a channel that is cooler than the surrounding plasma but still much hotter than a chemical flame. Samples to be analyzed are introduced into this central channel, usually as a mist of liquid formed by passing the liquid sample into a nebulizer.

To maximise plasma temperature (and hence ionisation efficiency) and stability, the sample should be introduced through the central tube with as little liquid (solvent load) as possible, and with consistent droplet sizes. A nebuliser can be used for liquid samples, followed by a spray chamber to remove larger droplets, or a desolvating nebuliser can be used to evaporate most of the solvent before it reaches the torch. Solid samples can also be introduced using laser ablation. The sample enters the central channel of the ICP, evaporates, molecules break apart, and then the constituent atoms ionise. At the temperatures prevailing in the plasma a significant proportion of the atoms of many chemical elements are ionized, each atom losing its most loosely bound electron to form a singly charged ion. The plasma temperature is selected to maximise ionisation efficiency for elements with a high first ionisation energy, while minimising second ionisation (double charging) for elements that have a low second ionisation energy.

Mass Spectrometry

For coupling to mass spectrometry, the ions from the plasma are extracted through a series of cones into a mass spectrometer, usually a quadrupole. The ions are separated on the basis of their mass-to-charge ratio and a detector receives an ion signal proportional to the concentration.

The concentration of a sample can be determined through calibration with certified reference material such as single or multi-element reference standards. ICP-MS also lends itself to quantitative determinations through isotope dilution, a single point method based on an isotopically enriched standard. In order to increase reproducibility and compensate errors but sensitivity variation, an internal standard can be added.

Other mass analyzers coupled to ICP systems include double focusing magnetic-electrostatic sector systems with both single and multiple collector, as well as time of flight systems (both axial and orthogonal accelerators have been used).

Applications

One of the largest volume uses for ICP-MS is in the medical and forensic field, specifically, toxicology. A physician may order a metal assay for a number of reasons, such as suspicion of heavy metal poisoning, metabolic concerns, and even hepatological issues.

Depending on the specific parameters unique to each patient's diagnostic plan, samples collected for analysis can range from whole blood, urine, plasma, serum, to even packed red blood cells. Another primary use for this instrument lies in the environmental field. Such applications include water testing for municipalities or private individuals all the way to soil, water and other material analysis for industrial purposes.

In recent years, industrial and biological monitoring has presented another major need for metal analysis via ICP-MS. Individuals working in factories where exposure to metals is likely and unavoidable, such as a battery factory, are required by their employer to have their blood or urine analyzed for metal toxicity on a regular basis. This monitoring has become a mandatory practice implemented by the U.S. Occupational Safety and Health Administration, in an effort to protect workers from their work environment and ensure proper rotation of work duties (i.e. rotating employees from a high exposure position to a low exposure position).

ICP-MS is also used widely in the geochemistry field for radiometric dating, in which it is used to analyze relative abundance of different isotopes, in particular uranium and lead. ICP-MS is more suitable for this application than the previously used thermal ionization mass spectrometry, as species with high ionization energy such as osmium and tungsten can be easily ionized. For high precision ratio work, multiple collector instruments are normally used to reduce the effect noise on the calculated ratios.

In the field of flow cytometry, a new technique uses ICP-MS to replace the traditional fluorochromes. Briefly, instead of labelling antibodies (or other biological probes) with fluorochromes, each antibody is labelled with a distinct combinations of lanthanides. When the sample of interest is analysed by ICP-MS in a specialised flow cytometer, each antibody can be identified and quantitated by virtue of a distinct ICP "footprint". In theory, hundreds of different biological probes can thus be analysed in an individual cell, at a rate of ca. 1,000 cells per second. Because elements are easily distinguished in ICP-MS, the problem of compensation in multiplex flow cytometry is effectively eliminated.

Laser ablation inductively coupled plasma mass spectrometry (LA-ICP-MS) is a powerful technique for the elemental analysis of a wide variety of materials encountered in forensic casework. (LA-ICP-MS) has already successfully been applied to applications in forensics, metals, glasses, soils, car paints, bones and teeth, printing inks, trace elemental, fingerprint, and paper. Among these, forensic glass analysis stands out as an application for which this technique has great utility to provide highly. Car hit and runs, burglaries, assaults, drive-by shootings and bombings such as these situations may cause glass fragments that could be used as evidence of association in glass transfer conditions. LA-ICP-MS is considered one of the best techniques for analysis of glass due to the short time for sample preparation and sample, small sample size of less than 250 nanograms. In addition there is no need for complex procedure and handling of dangerous materials that is used for digestion of the samples. This allows detecting major, minor and tracing elements with high level of precision and accuracy. There are set of properties that are used to measure glass sample such as physical and optical properties including color, thickness, density, refractive index (RI) and also, if necessary, elemental analysis can be conducted in order to enhance the value of an association.

Pharmaceutical Industry

In the pharmaceutical industry, ICP-MS is used for detecting inorganic impurities in pharmaceuticals and their ingredients. New and reduced maximum permitted exposure levels of heavy metals from dietary supplements, introduced in USP (United States Pharmacopeia) «*{232} Elemental Impurities—Limits*» and USP «*{232} Elemental Impurities—Procedures*», will increase the need for ICP-MS technology, where, previously, other analytic methods have been sufficient. Cosmetics, such as lipstick, recovered from a crime scene may provide valuable forensic information. Lipstick smears left on cigarette butts, glassware, clothing, bedding; napkins, paper, etc. may be valuable evidence. Lipstick recovered from clothing or skin may also indicate physical contact between individuals. Forensic analysis of recovered lipstick smear evidence can provide valuable information on the recent activities of a victim or suspect. Trace elemental analysis of lipstick smears could be used to complement existing visual comparative procedures to determine the lipstick brand and color.

Single Particle Inductively Coupled Plasma Mass Spectroscopy (SP ICP-MS) was designed for particle suspensions in 2000 by Claude Degueldre. He first tested this new methodology at the Forel Institute of the University of Geneva and presented this new analytical approach at the 'Colloid 2oo2' symposium during the spring 2002 meeting of the EMRS, and in the proceedings in 2003. This study presents the theory of SP ICP-MS and the results of tests carried out on clay particles (montmorillonite) as well as other suspensions of colloids.

This method was then tested on thorium dioxide nanoparticles by Degueldre & Favarger (2004), zirconium dioxide by Degueldre *et al* (2004) and gold nanoparticles, which are used as a substrate in nanopharmacy, and published by Degueldre *et al* (2006). Subsequently, the study of uranium dioxide nano- and micro-particles gave rise to a detailed publication, Ref. Degueldre *et al* (2006). Since 2010 the interest for SP ICP-MS has exploded.

Previous forensic techniques employed for the organic analysis of lipsticks by compositional comparison include thin layer chromatography (TLC), gas chromatography (GC), and high-performance liquid chromatography (HPLC). These methods provide useful information regarding the identification of lipsticks. However, they all require long sample preparation times and destroy the sample. Nondestructive techniques for the forensic analysis of lipstick smears include UV fluorescence observation combined with purge-and-trap gas chromatography, microspectrophotometry and scanning electron microscopy-energy dispersive spectroscopy (SEM-EDS), and Raman spectroscopy.

Metal Speciation

A growing trend in the world of elemental analysis has revolved around the speciation, or determination of oxidation state of certain metals such as chromium and arsenic. The toxicity of those elements varies with the oxidation state, so new regulations from food authorities requires speciation of some elements. One of the primary techniques to achieve this is to separate the chemical species with high-performance liquid chromatography (HPLC) or field flow fractionation (FFF) and then measure the concentrations with ICP-MS.

Quantification of Proteins and Biomolecules

There is an increasing trend of using ICP-MS as a tool in speciation analysis, which normally involves a front end chromatograph separation and an elemental selective detector, such as AAS and ICP-MS. For example, ICP-MS may be combined with size exclusion chromatography and preparative native PAGE for identifying and quantifying metalloproteins in biofluids. Also the phosphorylation status of proteins can be analyzed.

In 2007, a new type of protein tagging reagents called metal-coded affinity tags (MeCAT) were introduced to label proteins quantitatively with metals, especially lanthanides. The MeCAT labelling allows relative and absolute quantification of all kind of proteins or other biomolecules like peptides. MeCAT comprises a site-specific biomolecule tagging group with at least a strong chelate group which binds metals. The MeCAT labelled proteins can be accurately quantified by ICP-MS down to low attomol amount of analyte which is at least 2–3 orders of magnitude more sensitive than other mass spectrometry based quantification methods. By introducing several MeCAT labels to a biomolecule and further optimization of LC-ICP-MS detection limits in the zeptomol range are within the realm of possibility. By using different lanthanides MeCAT multiplexing can be used for pharmacokinetics of proteins and peptides or the analysis of the differential expression of proteins (proteomics) e.g. in biological fluids. Breakable PAGE SDS-PAGE (DPAGE, dissolvable PAGE), two-dimensional gel electrophoresis or chromatography is used for separation of MeCAT labelled proteins. Flow-injection ICP-MS analysis of protein bands or spots from DPAGE SDS-PAGE gels can be easily performed by dissolving the DPAGE gel after electrophoresis and staining of the gel. MeCAT labelled proteins are identified and relatively quantified on peptide level by MALDI-MS or ESI-MS.

Elemental Analysis

The ICP-MS allows determination of elements with atomic mass ranges 7 to 250 (Li to U), and sometimes higher. Some masses are prohibited such as 40 due to the abundance of argon in the sample. Other blocked regions may include mass 80 (due to the argon dimer), and mass 56 (due to ArO), the latter of which greatly hinders

Fe analysis unless the instrumentation is fitted with a reaction chamber. Such interferences can be reduced by using a high resolution ICP-MS (HR-ICP-MS) which uses two or more slits to constrict the beam and distinguish between nearby peaks. This comes at the cost of sensitivity. For example, distinguishing iron from argon requires a resolving power of about 10,000, which may reduce the iron sensitivity by around 99%.

A single collector ICP-MS may use a multiplier in pulse counting mode to amplify very low signals, an attenuation grid or a multiplier in analogue mode to detect medium signals, and a Faraday cup/bucket to detect larger signals. A multi-collector ICP-MS may have more than one of any of these, normally Faraday buckets which are much less expensive. With this combination, a dynamic range of 12 orders of magnitude, from 1 ppq to 100 ppm is possible.

ICP-MS is a method of choice for the determination of cadmium in biological samples. Unlike atomic absorption spectroscopy, which can only measure a single element at a time, ICP-MS has the capability to scan for all elements simultaneously. This allows rapid sample processing.

A simultaneous ICP-MS that can record the entire analytical spectrum from lithium to uranium in every analysis won the Silver Award at the 2010 Pittcon Editors' Awards. An ICP-MS may use multiple scan modes, each one striking a different balance between speed and precision. Using the magnet alone to scan is slow, due to hysteresis, but is precise. Electrostatic plates can be used in addition to the magnet to increase the speed, and this, combined with multiple collectors, can allow a scan of every element from Lithium 6 to Uranium Oxide 256 in less than a quarter of a second. For low detection limits, interfering species and high precision, the counting time can increase substantially. The rapid scanning, large dynamic range and large mass range is ideally suited to measuring multiple unknown concentrations and isotope ratios in samples that have had minimal preparation (an advantage over TIMS), for example seawater, urine, and digested whole rock samples. It also lends well to laser ablated rock samples, where the scanning rate is so quick that a real time plot of any number of isotopes is possible. This also allows easy spatial mapping of mineral grains.

Hardware

In terms of input and output, ICP-MS instrument consumes prepared sample material and translates it into mass-spectral data. Actual analytical procedure takes some time; after that time the instrument can be switched to work on the next sample. Series of such sample measurements requires the instrument to have plasma ignited, meanwhile a number of technical parameters has to be stable in order for the results obtained to have feasibly accurate and precise interpretation.

Maintaining the plasma requires a constant supply of carrier gas (usually, pure argon) and increased power consumption of the instrument. When these additional running costs are not considered justified, plasma and most of auxiliary systems can be turned off. In such standby mode only pumps are working to keep proper vacuum in mass-spectrometer. The constituents of ICP-MS instrument are designed to allow for reproducible and/or stable operation.

Sample Introduction

The first step in analysis is the introduction of the sample. This has been achieved in ICP-MS through a variety of means. The most common method is the use of *analytical nebulizers.* Nebulizer converts liquids into an aerosol, and that aerosol can then be swept into the plasma to create the ions. Nebulizers work best with simple liquid samples (i.e. solutions). However, there have been instances of their use with more complex materials like a slurry. Many varieties of nebulizers have been coupled to ICP-MS, including pneumatic, cross-flow, Babington, ultrasonic, and desolvating types. The aerosol generated is often treated to limit it to only smallest droplets, commonly by means of a Peltier cooled double pass or cyclonic spray chamber. Use of autosamplers makes this easier and faster, especially for routine work and large numbers of samples. A Desolvating Nebuliser (DSN) may also be used; this uses a long heated capillary, coated with a fluoropolymer membrane, to remove most of the solvent and reduce the load on the plasma. Matrix removal introduction systems are sometimes used for samples, such as seawater, where the species of interest are at trace levels, and are surrounded by much more abundant contaminants.

Laser ablation is another method. While being less common in the past, is rapidly becoming popular has been used as a means of sample introduction, thanks to increased ICP-MS scanning speeds. In this method, a pulsed UV laser is focused on the sample and creates a plume of ablated material which can be swept into the plasma. This allows geochemists to spacially map the isotope composition in cross-sections of rock samples, a tool which is lost if the rock is digested and introduced as a liquid sample. Lasers for this task are built to have highly controllable power outputs and uniform radial power distributions, to produce craters which are flat bottomed and of a chosen diameter and depth.

For both Laser Ablation and Desolvating Nebulisers, a small flow of Nitrogen may also be introduced into the Argon flow. Nitrogen exists as a dimer, so has more vibrational modes and is more efficient at receiving energy from the RF coil around the torch.

Other methods of sample introduction are also utilized. Electrothermal vaporization (ETV) and in torch vaporization (ITV) use hot surfaces (graphite or metal, generally) to vaporize samples for introduction. These can use very small amounts of liquids, solids, or slurries. Other methods like vapor generation are also known.

Plasma Torch

The plasma used in an ICP-MS is made by partially ionizing argon gas ($Ar \rightarrow Ar^+ + e^-$). The energy required for this reaction is obtained by pulsing an alternating electric current in load coil that surrounds the plasma torch with a flow of argon gas. After the sample is injected, the plasma's extreme temperature causes the sample to separate into individual atoms (atomization). Next, the plasma ionizes these atoms ($M \rightarrow M^+ + e^-$) so that they can be detected by the mass spectrometer.

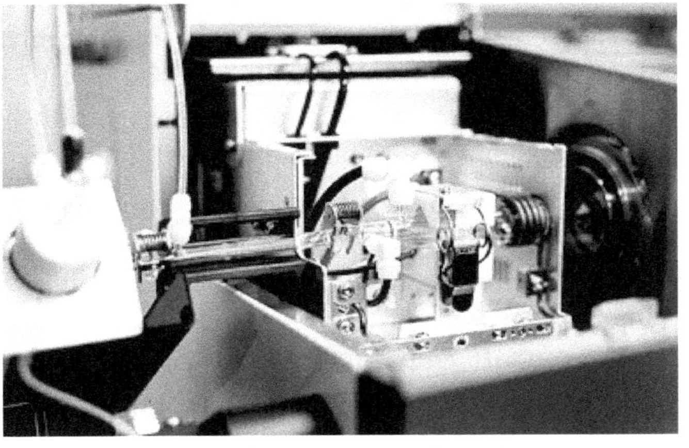

Fig 5: The Atomizer of an ICP

An inductively coupled plasma (ICP) for spectrometry is sustained in a torch that consists of three concentric tubes, usually made of quartz. The two major designs are the Fassel and Greenfield torches. The end of this torch is placed inside an induction coil supplied with a radio-frequency electric current. A flow of argon gas (usually 14 to 18 liters per minute) is introduced between the two outermost tubes of the torch and an electrical spark is applied for a short time to introduce free electrons into the gas stream. These electrons interact with the radio-frequency magnetic field of the induction coil and are accelerated first in one direction, then the other, as the field changes at high frequency (usually 27.12 MHz or 40 MHz). The accelerated electrons collide with argon atoms, and sometimes a collision causes an argon atom to part with one of its electrons. The released electron is in turn accelerated by the rapidly changing magnetic field. The process continues until the rate of release of new electrons in collisions is balanced by the rate of recombination of electrons with argon ions (atoms that have lost an electron). This produces a 'fireball' that consists mostly of argon atoms with a rather small fraction of free electrons and argon ions.

Advantage of Argon

Making the plasma from argon, instead of other gases, has several advantages. First, argon is abundant (in the atmosphere, as a result of the radioactive decay of potassium) and therefore cheaper than other noble gases. Argon also has a higher first ionization potential than all other elements except He, F, and Ne. Because of this high ionization energy, the reaction ($Ar^+ + e^- \rightarrow Ar$) is more energetically favorable than the reaction ($M^+ + e^- \rightarrow M$). This ensures that the sample remains ionized (as M^+) so that the mass spectrometer can detect it.

Argon can be purchased for use with the ICP-MS in either a refrigerated liquid or a gas form. However it is important to note that whichever form of argon purchased, it should have a guaranteed purity of 99.9% Argon at a minimum. It is important to determine which type of argon will be best suited for the specific situation. Liquid argon is typically cheaper and can be stored in a greater quantity as opposed to the gas form, which is more expensive and takes up more tank space.

If the instrument is in an environment where it gets infrequent use, then buying argon in the gas state will be most appropriate as it will be more than enough to suit smaller run times and gas in the cylinder will remain stable for longer periods of time, whereas liquid argon will suffer loss to the environment due to venting of the tank when stored over extended time frames.

However, if the ICP-MS is to be used routinely and is on and running for eight or more hours each day for several days a week, then going with liquid argon will be the most suitable. If there are to be multiple ICP-MS instruments running for long periods of time, then it will most likely be beneficial for the laboratory to install a bulk or micro bulk argon tank which will be maintained by a gas supply company, thus eliminating the need to change out tanks frequently as well as minimizing loss of argon that is left over in each used tank as well as down time for tank changeover.

Helium can be used either in place of, or mixed with, argon for plasma generation. Helium's higher first ionisation energy allows greater ionisation and therefore higher sensitivity for hard-to-ionise elements.

The use of pure helium also avoids argon-based interferences such as ArO. However, many of the interferences can be mitigated by use of a collision cell, and the greater cost of helium has prevented its use in commercial ICP-MS.

Transfer of Ions into Vacuum

The carrier gas is sent through the central channel and into the very hot plasma. The sample is then exposed to radio frequency which converts the gas into a plasma. The high temperature of the plasma is sufficient to cause a very large portion of the sample to form ions.

This fraction of ionization can approach 100% for some elements (e.g. sodium), but this is dependent on the ionization potential. A fraction of the formed ions passes through a ~1 mm hole (sampler cone) and then a ~0.4 mm hole (skimmer cone). The purpose of which is to allow a vacuum that is required by the mass spectrometer.

The vacuum is created and maintained by a series of pumps. The first stage is usually based on a roughing pump, most commonly a standard rotary vane pump. This removes most of the gas and typically reaches a pressure of around 133 Pa. Later stages have their vacuum generated by more powerful vacuum systems, most often turbomolecular pumps. Older instruments may have used oil diffusion pumps for high vacuum regions.

Ion Optics

Before mass separation, a beam of positive ions has to be extracted from the plasma and focused into the mass-analyzer. It is important to separate the ions from UV photons, energetic neutrals and from any solid particles that may have been carried into the instrument from the ICP. Traditionally, ICP-MS instruments have used transmitting ion lens arrangements for this purpose. Examples include the Einzel lens, the Barrel lens, Agilent's Omega Lens and Perkin-Elmer's Shadow Stop. Another approach is to use ion guides (quadrupoles, hexapoles, or octopoles) to guide the ions into mass analyzer along a path away from the trajectory of photons or neutral particles.

Yet another approach is Varian patented used by Analytik Jena ICP-MS 90 degrees reflecting parabolic "Ion Mirror" optics, which are claimed to provide more efficient ion transport into the mass-analyzer, resulting in better sensitivity and reduced background. Analytik Jena ICP-MS PQMS is the most sensitive instrument on the market.

A sector ICP-MS will commonly have four sections: an extraction acceleration region, steering lenses, an electrostatic sector and a magnetic sector. The first region takes ions from the plasma and accelerates them using a high voltage. The second uses may use a combination of parallel plates, rings, quadrupoles, hexapoles and octopoles to steer, shape and focus the beam so that the resulting peaks are symmetrical, flat topped and have high transmission.

The electrostatic sector may be before or after the magnetic sector depending on the particular instrument, and reduces the spread in kinetic energy caused by the plasma. This spread is particularly large for ICP-MS, being larger than Glow Discharge and much larger than TIMS. The geometry of the instrument is chosen so that the instrument the combined focal point of the electrostatic and magnetic sectors is at the collector, known as Double Focusing (or Double Focussing).

If the mass of interest has a low sensitivity and is just below a much larger peak, the low mass tail from this larger peak can intrude onto the mass of interest. A Retardation Filter might be used to reduce this tail. This sits near the collector, and applies a voltage equal but opposite to the accelerating voltage; any ions that have lost energy while flying around the instrument will be decelerated to rest by the filter.

Collision Reaction Cell and iCRC

The collision/reaction cell is used to remove interfering ions through ion/neutral reactions. Collision/reaction cells are known under several names. The dynamic reaction cell is located before the quadrupole in the ICP-MS device. The chamber has a quadrupole and can be filled with reaction (or collision) gases (ammonia, methane, oxygen or hydrogen), with one gas type at a time or a mixture of two of them, which reacts with the introduced sample, eliminating some of the interference.

The integrated Collisional Reaction Cell (iCRC) used by Analytik Jena ICP-MS is a mini-collision cell installed in front of the parabolic ion mirror optics that removes interfering ions by injecting a collisional gas (He), or a reactive gas (H_2), or a mixture of the two, directly into the plasma as it flows through the skimmer cone and/or the sampler cone. The iCRC removed interfering ions using a collisional kinetic energy discrimination (KED) phenomenon and chemical reactions with interfering ions similarly to traditionally used larger collision cells.

Routine Maintenance

As with any piece of instrumentation or equipment, there are many aspects of maintenance that need to be encompassed by daily, weekly and annual procedures. The frequency of maintenance is typically determined by the sample volume and cumulative run time that the instrument is subjected to.

One of the first things that should be carried out before the calibration of the ICP-MS is a sensitivity check and optimization. This ensures that the operator is aware of any possible issues with the instrument and if so, may address them before beginning a calibration. Typical indicators of sensitivity are Rhodium levels, Cerium/Oxide ratios and DI water blanks. One common standard practice is to measure a standard tuning solution provided by the ICP manufacturer every time the plasma torch is started. Then the instrument is auto-calibrated for optimum sensitivity and the operator obtains a report providing certain parameters such as sensitivity, mass resolution and estimated amount of oxidized species and double-positive charged species.

One of the most frequent forms of routine maintenance is replacing sample and waste tubing on the peristaltic pump, as these tubes can get worn fairly quickly resulting in holes and clogs in the sample line, resulting in skewed results. Other parts that will need regular cleaning and/or replacing are sample tips, nebulizer tips, sample cones, skimmer cones, injector tubes, torches and lenses. It may also be necessary to change the oil in the interface roughing pump as well as the vacuum backing pump, depending on the workload put on the instrument.

Sample Preparation

For most clinical methods using ICP-MS, there is a relatively simple and quick sample prep process. The main component to the sample is an internal standard, which also serves as the diluent. This internal standard consists primarily of deionized water, with nitric or hydrochloric acid, and Indium and/or Gallium. The addition of volatile acids allows for the sample to decompose into its gaseous components in the plasma which minimizes the ability for concentrated salts and solvent loads to clog the cones and contaminate the instrument. Depending on the sample type, usually 5 mL of the internal standard is added to a test tube along with 10–500 microliters of sample. This mixture is then vortexed for several seconds or until mixed well and then loaded onto the autosampler tray. For other applications that may involve very viscous samples or samples that have particulate matter, a process known as sample digestion may have to be carried out, before it can be pipetted and analyzed. This adds an extra first step to the above process, and therefore makes the sample prep more lengthy.

5.3.6 Laser-Induced Breakdown Spectroscopy (LIBS)

Laser-induced breakdown spectroscopy (LIBS) is a type of atomic emission spectroscopy which uses a highly energetic laser pulse as the excitation source. The laser is focused to form a plasma, which atomizes and excites samples. The formation of the plasma only begins when the focused laser achieves a certain threshold for optical breakdown, which generally depends on the environment and the target material.

2000s Developments

From 2000 to 2010, the U.S. Army Research Laboratory (ARL) researched potential extensions to LIBS technology, which focused on hazardous material detection. Applications investigated at ARL included the standoff detection of explosive residues and other hazardous materials, plastic landmine discrimination, and material characterization of various metal alloys and polymers. Results presented by ARL suggest that LIBS may be able to discriminate between energetic and non-energetic materials.

Research

Broadband high-resolution spectrometers were developed in 2000 and commercialized in 2003. Designed for material analysis, the spectrometer allowed the LIBS system to be sensitive to chemical elements in low concentration. ARL LIBS applications studied from 2000 to 2010 included:

- Tested for detection of Halon alternative agents.
- Tested a field-portable LIBS system for the detection of lead in soil and paint.
- Studied the spectral emission of aluminum and aluminum oxides from bulk aluminum in different bath gases.
- Performed kinetic modeling of LIBS plumes.
- Demonstrated the detection and discrimination of geological materials, plastic landmines, explosives, and chemical and biological warfare agent surrogates.

ARL LIBS prototypes studied during this period included:

- Laboratory LIBS setup.
- Commercial LIBS system.
- Man-portable LIBS device.
- Standoff LIBS system developed for 100+ m detection and discriminate on of explosive residues.

2010s Developments

LIBS is one of several analytical techniques that can be deployed in the field as opposed to pure laboratory techniques e.g. spark OES. As of 2015, recent research on LIBS focuses on compact and (man-)portable systems. Some industrial applications of LIBS include the detection of material mix-ups, analysis of inclusions in steel, analysis of slags in secondary metallurgy, analysis of combustion processes, and high-speed identification of scrap pieces for material-specific recycling tasks. Armed with data analysis techniques, this technique is being extended to pharmaceutical samples.

LIBS Using Short Laser Pulses

Following multiphoton or tunnel ionization the electron is being accelerated by inverse Bremsstrahlung and can collide with the nearby molecules and generate new electrons through collisions. If the pulse duration is long, the newly ionized electrons can be accelerated and eventually avalanche or cascade ionization follows. Once the density of the electrons reaches a critical value, breakdown occurs and high density plasma is created which has no memory of the laser pulse. So, the criterion for the shortness of a pulse in dense media is as follows: A pulse interacting with a dense matter is considered to be short if during the interaction the threshold for the avalanche ionization is not reached. At the first glance this definition may appear to be too limiting. Fortunately, due to the delicately balanced behavior of the pulses in dense media, the threshold cannot be reached easily. The phenomenon responsible for the balance is the intensity clamping through the onset of filamentation process during the propagation of strong laser pulses in dense media.

A potentially important development to LIBS involves the use of a short laser pulse as a spectroscopic source. In this method, a plasma column is created as a result of focusing ultrafast laser pulses in a gas. The self-luminous plasma is far superior in terms of low level of continuum and also smaller line broadening. This is attributed to the lower density of the plasma in the case of short laser pulses due to the defocusing effects which limits the intensity of the pulse in the interaction region and thus prevents further multiphoton/tunnel ionization of the gas.

LIBS for Food Analysis

Recently, LIBS has been investigated as a fast, micro-destructive food analysis tool. It is considered a potential analytical tool for qualitative and quantitative chemical analysis, making it suitable as a PAT (Process Analytical Technology) or portable tool. Milk, bakery products, tea, vegetable oils, water, cereals, flour, potatoes, palm date and different types of meat have been analyzed using LIBS. Few studies have shown its potential as an adulteration detection tool for certain foods. LIBS has also been evaluated as a promising elemental imaging technique in meat.

In 2019, researchers of the University of York and of the Liverpool John Moores University employed LIBS for studying 12 European oysters (*Ostrea edulis*, Linnaeus, 1758) from the Late Mesolithic shell midden at Conors Island (Republic of Ireland). The results highlighted the applicability of LIBS to determine prehistoric seasonality practices as well as biological age and growth at an improved rate and reduced cost than was previously achievable.

5.4 APPLICATIONS OF ADVANCED SPECTROSCOPY IN FORENSIC CHEMISTRY

The applications of advanced spectroscopy in forensic chemistry are vast, ranging from drug analysis and toxicology to trace evidence and environmental crime investigation. Below are some of the key areas where advanced spectroscopic techniques play a crucial role in forensic investigations.

5.4.1 Drug Analysis

Advanced spectroscopic techniques, such as FTIR, Raman, and NMR spectroscopy, are widely used in the analysis of controlled substances. These techniques allow forensic chemists to identify and quantify drugs in seized materials or biological samples, helping to establish the presence of illegal substances and determine their purity.

Spectroscopy is particularly valuable in cases involving novel psychoactive substances (NPS), which are often chemically similar to known drugs but differ slightly in structure. Advanced spectroscopic methods can provide detailed information about the molecular structure of these substances, helping to classify them and assess their legal status.

5.4.2 Toxicology

In forensic toxicology, advanced spectroscopic techniques are used to detect and quantify toxic substances in biological samples. ICP-MS, for example, is commonly used to analyze blood, urine, or tissue samples for the presence of toxic metals or other harmful elements. NMR spectroscopy can provide detailed information about the metabolites of drugs or poisons, helping to determine the cause of death in poisoning cases.

5.4.3 Trace Evidence Analysis

Trace evidence, such as fibers, paints, and gunshot residue, can be critical in linking suspects to crime scenes. Advanced spectroscopic techniques, such as XRF, LIBS, and Raman spectroscopy, allow forensic scientists to analyze the chemical composition of trace evidence with high sensitivity and specificity. These methods can provide valuable information about the source of the evidence, helping to establish connections between suspects, victims, and crime scenes.

5.5 CONCLUSION

Advanced spectroscopy has revolutionized forensic chemistry, providing forensic scientists with powerful tools for analyzing and identifying a wide range of materials. Techniques such as FTIR, Raman, NMR, XRF, ICP-MS, and LIBS offer unparalleled sensitivity, resolution, and versatility, allowing forensic chemists to analyze complex samples and detect trace amounts of evidence. The non-destructive nature of many spectroscopic methods also ensures that evidence can be preserved for further analysis. As forensic science continues to evolve, advanced spectroscopy will remain at the forefront of forensic investigations, contributing to more accurate and reliable results in the pursuit of justice.

CHAPTER 6
CHROMATOGRAPHY

6.1 UNDERSTANDING THE CHROMATOGRAPHY

In chemical analysis, chromatography is a laboratory technique for the separation of a mixture into its components. The mixture is dissolved in a fluid solvent (gas or liquid) called the *mobile phase*, which carries it through a system (a column, a capillary tube, a plate, or a sheet) on which a material called the *stationary phase* is fixed. Because the different constituents of the mixture tend to have different affinities for the stationary phase and are retained for different lengths of time depending on their interactions with its surface sites, the constituents travel at different apparent velocities in the mobile fluid, causing them to separate. The separation is based on the differential partitioning between the mobile and the stationary phases. Subtle differences in a compound's partition coefficient result in differential retention on the stationary phase and thus affect the separation.

Chromatography may be *preparative* or *analytical*. The purpose of preparative chromatography is to separate the components of a mixture for later use, and is thus a form of purification. This process is associated with higher costs due to its mode of production. Analytical chromatography is done normally with smaller amounts of material and is for establishing the presence or measuring the relative proportions of analytes in a mixture. The two types are not mutually exclusive.

6.1.1 History

Chromatography was first devised at the University of Kazan by the Italian-born Russian scientist Mikhail Tsvet in 1900. He developed the technique and coined the term *chromatography* in the first decade of the 20th century, primarily for the separation of plant pigments such as chlorophyll, carotenes, and xanthophylls. Since these components separate in bands of different colors (green, orange, and yellow, respectively) they directly inspired the name of the technique. New types of chromatography developed during the 1930s and 1940s made the technique useful for many separation processes.

Chromatography technique developed substantially as a result of the work of Archer John Porter Martin and Richard Laurence Millington Synge during the 1940s and 1950s, for which they won the 1952 Nobel Prize in Chemistry. They established the principles and basic techniques of partition chromatography, and their work encouraged the rapid development of several chromatographic methods: paper chromatography, gas chromatography, and what would become known as high-performance liquid chromatography. Since then, the technology has advanced rapidly. Researchers found that the main principles of Tsvet's chromatography could be applied in many different ways, resulting in the different varieties of chromatography described below. Advances are continually improving the technical performance of chromatography, allowing the separation of increasingly similar molecules.

6.1.2 Terms

- **Analyte:** The substance to be separated during chromatography. It is also normally what is needed from the mixture.
- **Analytical Chromatography:** The use of chromatography to determine the existence and possibly also the concentration of analyte(s) in a sample.
- **Bonded Phase:** A stationary phase that is covalently bonded to the support particles or to the inside wall of the column tubing.
- **Chromatogram:** The visual output of the chromatograph. In the case of an optimal separation, different peaks or patterns on the chromatogram correspond to different components of the separated mixture. Plotted on the x-axis is the retention time and plotted on the y-axis a signal (for example obtained by a spectrophotometer, mass spectrometer or a variety of other detectors) corresponding to the response created by the analytes exiting the system. In the case of an optimal system the signal is proportional to the concentration of the specific analyte separated.

- **Chromatograph:** An instrument that enables a sophisticated separation, e.g. gas chromatographic or liquid chromatographic separation.
- **Chromatography:** A physical method of separation that distributes components to separate between two phases, one stationary (stationary phase), the other (the mobile phase) moving in a definite direction.
- **Eluent (sometimes Spelled Eluant):** The solvent or solvent fixure used in elution chromatography and is synonymous with *mobile phase.*

- **Eluate:** The mixture of *solute* and *solvent* exiting the column.
- **Effluent:** The stream flowing out of a chromatographic column. In practise, it is used synonymously with *eluate*, but the term more precisely refers to the stream independent of separation taking place.
- **Eluite:** A more precise term for *solute* or *analyte*. It is a sample component leaving the chromatographic column.
- **Eluotropic Series:** A list of solvents ranked according to their eluting power.
- **Immobilized Phase:** A stationary phase that is immobilized on the support particles, or on the inner wall of the column tubing.
- **Mobile Phase:** The phase that moves in a definite direction. It may be a liquid (LC and capillary electrochromatography, CEC), a gas (GC), or a supercritical fluid (supercritical-fluid chromatography, SFC). The mobile phase consists of the sample being separated/analyzed and the solvent that moves the sample through the column. In the case of HPLC the mobile phase consists of a non-polar solvent(s) such as hexane in normal phase or a polar solvent such as methanol in reverse phase chromatography and the sample being separated. The mobile phase moves through the chromatography column (the stationary phase) where the sample interacts with the stationary phase and is separated.
- **Preparative Chromatography:** The use of chromatography to purify sufficient quantities of a substance for further use, rather than analysis.
- **Retention Time:** The characteristic time it takes for a particular analyte to pass through the system (from the column inlet to the detector) under set conditions.
- **Sample:** The matter analyzed in chromatography. It may consist of a single component or it may be a mixture of components. When the sample is treated in the course of an analysis, the phase or the phases containing the analytes of interest is/are referred to as the sample whereas everything out of interest separated from the sample before or in the course of the analysis is referred to as *waste*.
- **Solute:** The sample components in partition chromatography.
- **Solvent:** Any substance capable of solubilizing another substance, and especially the liquid mobile phase in liquid chromatography.
- **Stationary Phase:** The substance fixed in place for the chromatography procedure. Examples include the silica layer in thin-layer chromatography

- **Detector:** The instrument used for qualitative and quantitative detection of analytes after separation.

Chromatography is based on the concept of partition coefficient. Any solute partitions between two immiscible solvents. When one make one solvent immobile (by adsorption on a solid support matrix) and another mobile it results in most common applications of chromatography. If the matrix support, or stationary phase, is polar (e.g., cellulose, silica etc.) it is forward phase chromatography. Otherwise this technique is known as reversed phase, where a non-polar stationary phase (e.g., non-polar derivative of C-18) is used.

6.1.3 Techniques by Chromatographic Bed Shape

Column Chromatography

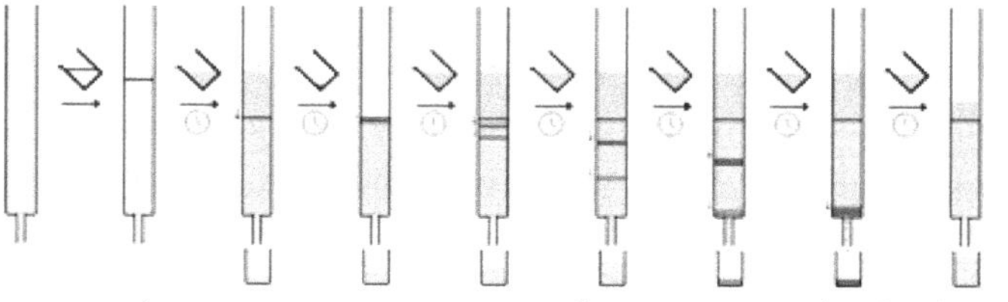

Column chromatography is a separation technique in which the stationary bed is within a tube. The particles of the solid stationary phase or the support coated with a liquid stationary phase may fill the whole inside volume of the tube (packed column) or be concentrated on or along the inside tube wall leaving an open, unrestricted path for the mobile phase in the middle part of the tube (open tubular column). Differences in rates of movement through the medium are calculated to different retention times of the sample. In 1978, W. Clark Still introduced a modified version of column chromatography called *flash column chromatography* (flash). The technique is very similar to the traditional column chromatography, except that the solvent is driven through the column by applying positive pressure. This allowed most separations to be performed in less than 20 minutes, with improved separations compared to the old method. Modern flash chromatography systems are sold as pre-packed plastic cartridges, and the solvent is pumped through the cartridge. Systems may also be linked with detectors and fraction collectors providing automation. The introduction of gradient pumps resulted in quicker separations and less solvent usage.

In expanded bed adsorption, a fluidized bed is used, rather than a solid phase made by a packed bed. This allows omission of initial clearing steps such as centrifugation and filtration, for culture broths or slurries of broken cells.

Phosphocellulose chromatography utilizes the binding affinity of many DNA-binding proteins for phosphocellulose. The stronger a protein's interaction with DNA, the higher the salt concentration needed to elute that protein.

Planar Chromatography

Planar chromatography is a separation technique in which the stationary phase is present as or on a plane. The plane can be a paper, serving as such or impregnated by a substance as the stationary bed (paper chromatography) or a layer of solid particles spread on a support such as a glass plate (thin-layer chromatography). Different compounds in the sample mixture travel different distances according to how strongly they interact with the stationary phase as compared to the mobile phase. The specific Retention factor (R_f) of each chemical can be used to aid in the identification of an unknown substance.

Paper Chromatography

Paper chromatography is a technique that involves placing a small dot or line of sample solution onto a strip of *chromatography paper*. The paper is placed in a container with a shallow layer of solvent and sealed. As the solvent rises through the paper, it meets the sample mixture, which starts to travel up the paper with the solvent. This paper is made of cellulose, a polar substance, and the compounds within the mixture travel further if they are less polar. More polar substances bond with the cellulose paper more quickly, and therefore do not travel as far.

Thin-Layer Chromatography (TLC)

Thin-layer chromatography (TLC) is a widely employed laboratory technique used to separate different biochemicals on the basis of their relative attractions to the stationary and mobile phases. It is similar to paper chromatography. However, instead of using a stationary phase of paper, it involves a stationary phase of a thin layer of adsorbent like silica gel, alumina, or cellulose on a flat, inert substrate. TLC is very versatile; multiple samples can be separated simultaneously on the same layer, making it very useful for screening applications such as testing drug levels and water purity.

Possibility of cross-contamination is low since each separation is performed on a new layer. Compared to paper, it has the advantage of faster runs, better separations, better quantitative analysis, and the choice between different adsorbents. For even better resolution and faster separation that utilizes less solvent, high-performance TLC can be used. An older popular use had been to differentiate chromosomes by observing distance in gel (separation of was a separate step).

6.1.4 Displacement Chromatography

The basic principle of displacement chromatography is: A molecule with a high affinity for the chromatography matrix (the displacer) competes effectively for binding sites, and thus displaces all molecules with lesser affinities. There are distinct differences between displacement and elution chromatography. In elution mode, substances typically emerge from a column in narrow, Gaussian peaks. Wide separation of peaks, preferably to baseline, is desired for maximum purification. The speed at which any component of a mixture travels down the column in elution mode depends on many factors.

But for two substances to travel at different speeds, and thereby be resolved, there must be substantial differences in some interaction between the biomolecules and the chromatography matrix. Operating parameters are adjusted to maximize the effect of this difference. In many cases, baseline separation of the peaks can be achieved only with gradient elution and low column loadings. Thus, two drawbacks to elution mode chromatography, especially at the preparative scale, are operational complexity, due to gradient solvent pumping, and low throughput, due to low column loadings. Displacement chromatography has advantages over elution chromatography in that components are resolved into consecutive zones of pure substances rather than "peaks". Because the process takes advantage of the nonlinearity of the isotherms, a larger column feed can be separated on a given column with the purified components recovered at significantly higher concentrations.

6.1.5 Techniques by Physical State of Mobile Phase

Gas Chromatography

Gas chromatography (GC), also sometimes known as gas-liquid chromatography, (GLC), is a separation technique in which the mobile phase is a gas. Gas chromatographic separation is always carried out in a column, which is typically "packed" or "capillary". Packed columns are the routine work horses of gas chromatography, being cheaper and easier to use and often giving adequate performance. Capillary columns generally give far superior resolution and although more expensive are becoming widely used, especially for complex mixtures.

Further, capillary columns can be split into three classes: porous layer open tubular (PLOT), wall-coated open tubular (WCOT) and support-coated open tubular (SCOT) columns. PLOT columns are unique in a way that the stationary phase is adsorbed to the column walls, while WCOT columns have a stationary phase that is chemically bonded to the walls. SCOT columns are in a way the combination of the two types mentioned in a way that they have support particles adhered to column walls, but those particles have liquid phase chemically bonded onto them. Both types of column are made from non-adsorbent and chemically inert materials. Stainless steel and glass are the usual materials for packed columns and quartz or fused silica for capillary columns.

Gas chromatography is based on a partition equilibrium of analyte between a solid or viscous liquid stationary phase (often a liquid silicone-based material) and a mobile gas (most often helium). The stationary phase is adhered to the inside of a small-diameter (commonly 0.53 – 0.18mm inside diameter) glass or fused-silica tube (a capillary column) or a solid matrix inside a larger metal tube (a packed column). It is widely used in analytical chemistry; though the high temperatures used in GC make it unsuitable for high molecular weight biopolymers or proteins (heat denatures them), frequently encountered in biochemistry, it is well suited for use in the petrochemical, environmental monitoring and remediation, and industrial chemical fields. It is also used extensively in chemistry research.

Liquid Chromatography

Liquid chromatography (LC) is a separation technique in which the mobile phase is a liquid. It can be carried out either in a column or a plane. Present day liquid chromatography that generally utilizes very small packing particles and a relatively high pressure is referred to as high-performance liquid chromatography.

In HPLC the sample is forced by a liquid at high pressure (the mobile phase) through a column that is packed with a stationary phase composed of irregularly or spherically shaped particles, a porous monolithic layer, or a porous membrane. Monoliths are "sponge-like chromatographic media" and are made up of an unending block of organic or inorganic parts. HPLC is historically divided into two different sub-classes based on the polarity of the mobile and stationary phases. Methods in which the stationary phase is more polar than the mobile phase (e.g., toluene as the mobile phase, silica as the stationary phase) are termed normal phase liquid chromatography (NPLC) and the opposite (e.g., water-methanol mixture as the mobile phase and C18 (octadecylsilyl) as the stationary phase) is termed reversed phase liquid chromatography (RPLC).

Supercritical Fluid Chromatography

Supercritical fluid chromatography is a separation technique in which the mobile phase is a fluid above and relatively close to its critical temperature and pressure. Specific techniques under this broad heading are listed below.

6.1.6 Affinity Chromatography

Affinity chromatography is based on selective non-covalent interaction between an analyte and specific molecules. It is very specific, but not very robust. It is often used in biochemistry in the purification of proteins bound to tags. These fusion proteins are labeled with compounds such as His-tags, biotin or antigens, which bind to the stationary phase specifically. After purification, these tags are usually removed and the pure protein is obtained.

Affinity chromatography often utilizes a biomolecule's affinity for the cations of a metal (Zn, Cu, Fe, etc.). Columns are often manually prepared and could be designed specifically for the proteins of interest. Traditional affinity columns are used as a preparative step to flush out unwanted biomolecules, or as a primary step in analyzing a protein with unknown physical properties.

However, liquid chromatography techniques exist that do utilize affinity chromatography properties. Immobilized metal affinity chromatography (IMAC) is useful to separate the aforementioned molecules based on the relative affinity for the metal. Often these columns can be loaded with different metals to create a column with a targeted affinity.

6.1.7 Techniques by Separation Mechanism

Ion Exchange Chromatography

Ion exchange chromatography (usually referred to as ion chromatography) uses an ion exchange mechanism to separate analytes based on their respective charges. It is usually performed in columns but can also be useful in planar mode. Ion exchange chromatography uses a charged stationary phase to separate charged compounds including anions, cations, amino acids, peptides, and proteins. In conventional methods the stationary phase is an ion-exchange resin that carries charged functional groups that interact with oppositely charged groups of the compound to retain. There are two types of ion exchange chromatography: Cation-Exchange and Anion-Exchange. In the Cation-Exchange Chromatography the stationary phase has negative charge and the exchangeable ion is a cation, whereas, in the Anion-Exchange Chromatography the stationary phase has positive charge and the exchangeable ion is an anion. Ion exchange chromatography is commonly used to purify proteins using FPLC.

Size-Exclusion Chromatography

Size-exclusion chromatography (SEC) is also known as *gel permeation chromatography* (GPC) or *gel filtration chromatography* and separates molecules according to their size (or more accurately according to their hydrodynamic diameter or hydrodynamic volume). Smaller molecules are able to enter the pores of the media and, therefore, molecules are trapped and removed from the flow of the mobile phase. The average residence time in the pores depends upon the effective size of the analyte molecules. However, molecules that are larger than the average pore size of the packing are excluded and thus suffer essentially no retention; such species are the first to be eluted. It is generally a low-resolution chromatography technique and thus it is often reserved for the final, "polishing" step of a purification. It is also useful for determining the tertiary structure and quaternary structure of purified proteins, especially since it can be carried out under native solution conditions.

Expanded Bed Adsorption Chromatographic Separation

An expanded bed chromatographic adsorption (EBA) column for a biochemical separation process comprises a pressure equalization liquid distributor having a self-cleaning function below a porous blocking sieve plate at the bottom of the expanded bed, an upper part nozzle assembly having a backflush cleaning function at the top of the expanded bed, a better distribution of the feedstock liquor added into the expanded bed ensuring that the fluid passed through the expanded bed layer displays a state of piston flow. The expanded bed layer displays a state of piston flow. The expanded bed chromatographic separation column has advantages of increasing the separation efficiency of the expanded bed.

Expanded-bed adsorption (EBA) chromatography is a convenient and effective technique for the capture of proteins directly from unclarified crude sample. In EBA chromatography, the settled bed is first expanded by upward flow of equilibration buffer. The crude feed, which is a mixture of soluble proteins, contaminants, cells, and cell debris, is then passed upward through the expanded bed. Target proteins are captured on the adsorbent, while particulates and contaminants pass through. A change to elution buffer while maintaining upward flow results in desorption of the target protein in expanded-bed mode. Alternatively, if the flow is reversed, the adsorbed particles will quickly settle and the proteins can be desorbed by an elution buffer. The mode used for elution (expanded-bed versus settled-bed) depends on the characteristics of the feed. After elution, the adsorbent is cleaned with a predefined cleaning-in-place (CIP) solution, with cleaning followed by either column regeneration (for further use) or storage.

6.1.8 Special Techniques

Reversed-Phase Chromatography

Reversed-phase chromatography (RPC) is any liquid chromatography procedure in which the mobile phase is significantly more polar than the stationary phase. It is so named because in normal-phase liquid chromatography, the mobile phase is significantly less polar than the stationary phase. Hydrophobic molecules in the mobile phase tend to adsorb to the relatively hydrophobic stationary phase. Hydrophilic molecules in the mobile phase will tend to elute first. Separating columns typically comprise a C8 or C18 carbon-chain bonded to a silica particle substrate.

Hydrophobic Interaction Chromatography

Hydrophobic Interaction Chromatography (HIC) is a purification and analytical technique that separates analytes, such as proteins, based on hydrophobic interactions between that analyte and the chromatographic matrix. It can provide a non-denaturing orthogonal approach to reversed phase separation, preserving native structures and potentially protein activity. In hydrophobic interaction chromatography, the matrix material is lightly substituted with hydrophobic groups. These groups can range from methyl, ethyl, propyl, butyl, octyl, or phenyl groups. At high salt concentrations, non-polar sidechains on the surface on proteins "interact" with the hydrophobic groups; that is, both types of groups are excluded by the polar solvent (hydrophobic effects are augmented by increased ionic strength). Thus, the sample is applied to the column in a buffer which is highly polar, which drives an association of hydrophobic patches on the analyte with the stationary phase. The eluent is typically an aqueous buffer with decreasing salt concentrations, increasing concentrations of detergent (which disrupts hydrophobic interactions), or changes in pH. Of critical importance is the type of salt used, with more kosmotropic salts as defined by the Hofmeister series providing the most water structuring around the molecule and resulting hydrophobic pressure. Ammonium sulfate is frequently used for this purpose. The addition of organic solvents or other less polar constituents may assist in improving resolution.

In general, Hydrophobic Interaction Chromatography (HIC) is advantageous if the sample is sensitive to pH change or harsh solvents typically used in other types of chromatography but not high salt concentrations. Commonly, it is the amount of salt in the buffer which is varied. In 2012, Müller and Franzreb described the effects of temperature on HIC using Bovine Serum Albumin (BSA) with four different types of hydrophobic resin. The study altered temperature as to effect the binding affinity of BSA onto the matrix. It was concluded that cycling temperature from 40 to 10 degrees Celsius would not be adequate to effectively wash all BSA from the matrix but could be very effective if the column would only be used a few times. Using temperature to effect change allows labs to cut costs on buying salt and saves money.

If high salt concentrations along with temperature fluctuations want to be avoided one can use a more hydrophobic to compete with one's sample to elute it. This so-called salt independent method of HIC showed a direct isolation of Human Immunoglobulin G (IgG) from serum with satisfactory yield and used β-cyclodextrin as a competitor to displace IgG from the matrix. This largely opens up the possibility of using HIC with samples which are salt sensitive as we know high salt concentrations precipitate proteins.

Hydrodynamic Chromatography

Hydrodynamic chromatography (HDC) is derived from the observed phenomenon that large droplets move faster than small ones. In a column, this happens because the center of mass of larger droplets is prevented from being as close to the sides of the column as smaller droplets because of their larger overall size. Larger droplets will elute first from the middle of the column while smaller droplets stick to the sides of the column and elute last. This form of chromatography is useful for separating analytes by molar mass (or molecular mass), size, shape, and structure when used in conjunction with light scattering detectors, viscometers, and refractometers. The two main types of HDC are open tube and packed column. Open tube offers rapid separation times for small particles, whereas packed column HDC can increase resolution and is better suited for particles with an average molecular mass larger than 10 5 10^5 daltons. HDC differs from other types of chromatography because the separation only takes place in the interstitial volume, which is the volume surrounding and in between particles in a packed column.

HDC shares the same order of elution as Size Exclusion Chromatography (SEC) but the two processes still vary in many ways. In a study comparing the two types of separation, Isenberg, Brewer, Côté, and Striegel use both methods for polysaccharide characterization and conclude that HDC coupled with multiangle light scattering (MALS) achieves more accurate molar mass distribution when compared to off-line MALS than SEC in significantly less time. This is largely due to SEC being a more destructive technique because of the pores in the column degrading the analyte during separation, which tends to impact the mass distribution. However, the main disadvantage of HDC is low resolution of analyte peaks, which makes SEC a more viable option when used with chemicals that are not easily degradable and where rapid elution is not important.

HDC plays an especially important role in the field of microfluidics. The first successful apparatus for HDC-on-a-chip system was proposed by Chmela, et al. in 2002. Their design was able to achieve separations using an 80 mm long channel on the timescale of 3 minutes for particles with diameters ranging from 26 to 110 nm, but the authors expressed a need to improve the retention and dispersion parameters. In a 2010 publication by Jellema, Markesteijn, Westerweel, and Verpoorte, implementing HDC with a recirculating bidirectional flow resulted in high resolution, size based separation with only a 3 mm long channel. Having such a short channel and high resolution was viewed as especially impressive considering that previous studies used channels that were 80 mm in length. For a biological application, in 2007, Huh, et al. proposed a microfluidic sorting device based on HDC and gravity, which was useful for preventing potentially dangerous particles with diameter larger than 6 microns from entering the bloodstream when injecting contrast agents in ultrasounds. This study also made advances for environmental sustainability in microfluidics due to the lack of outside electronics driving the flow, which came as an advantage of using a gravity based device.

Two-Dimensional Chromatography

In some cases, the selectivity provided by the use of one column can be insufficient to provide resolution of analytes in complex samples. Two-dimensional chromatography aims to increase the resolution of these peaks by using a second column with different physico-chemical (chemical classification) properties.

Since the mechanism of retention on this new solid support is different from the first dimensional separation, it can be possible to separate compounds by two-dimensional chromatography that are indistinguishable by one-dimensional chromatography. Furthermore, the separation on the second dimension occurs faster than the first dimension. An example of a TDC separation is where the sample is spotted at one corner of a square plate, developed, air-dried, then rotated by 90° and usually redeveloped in a second solvent system.

Two-dimensional chromatography can be applied to GC or LC separations. The heart-cutting approach selects a specific region of interest on the first dimension for separation, and the comprehensive approach uses all analytes in the second-dimension separation.

Simulated Moving-Bed Chromatography

The simulated moving bed (SMB) technique is a variant of high performance liquid chromatography; it is used to separate particles and/or chemical compounds that would be difficult or impossible to resolve otherwise. This increased separation is brought about by a valve-and-column arrangement that is used to lengthen the stationary phase indefinitely. In the moving bed technique of preparative chromatography the feed entry and the analyte recovery are simultaneous and continuous, but because of practical difficulties with a continuously moving bed, simulated moving bed technique was proposed. In the simulated moving bed technique instead of moving the bed, the sample inlet and the analyte exit positions are moved continuously, giving the impression of a moving bed. True moving bed chromatography (TMBC) is only a theoretical concept. Its simulation, SMBC is achieved by the use of a multiplicity of columns in series and a complex valve arrangement. This valve arrangement provides for sample and solvent feed and analyte and waste takeoff at appropriate locations of any column, whereby it allows switching at regular intervals the sample entry in one direction, the solvent entry in the opposite direction, whilst changing the analyte and waste takeoff positions appropriately as well.

Pyrolysis Gas Chromatography

Pyrolysis–gas chromatography–mass spectrometry is a method of chemical analysis in which the sample is heated to decomposition to produce smaller molecules that are separated by gas chromatography and detected using mass spectrometry.

Pyrolysis is the thermal decomposition of materials in an inert atmosphere or a vacuum. The sample is put into direct contact with a platinum wire, or placed in a quartz sample tube, and rapidly heated to 600–1000 °C. Depending on the application even higher temperatures are used. Three different heating techniques are used in actual pyrolyzers: Isothermal furnace, inductive heating (Curie point filament), and resistive heating using platinum filaments. Large molecules cleave at their weakest points and produce smaller, more volatile fragments. These fragments can be separated by gas chromatography. Pyrolysis GC chromatograms are typically complex because a wide range of different decomposition products is formed. The data can either be used as fingerprints to prove material identity or the GC/MS data is used to identify individual fragments to obtain structural information. To increase the volatility of polar fragments, various methylating reagents can be added to a sample before pyrolysis.

Besides the usage of dedicated pyrolyzers, pyrolysis GC of solid and liquid samples can be performed directly inside Programmable Temperature Vaporizer (PTV) injectors that provide quick heating (up to 30 °C/s) and high maximum temperatures of 600–650 °C. This is sufficient for some pyrolysis applications. The main advantage is that no dedicated instrument has to be purchased and pyrolysis can be performed as part of routine GC analysis. In this case, quartz GC inlet liners have to be used. Quantitative data can be acquired, and good results of derivatization inside the PTV injector are published as well.

Fast Protein Liquid Chromatography

Fast protein liquid chromatography (FPLC), is a form of liquid chromatography that is often used to analyze or purify mixtures of proteins. As in other forms of chromatography, separation is possible because the different components of a mixture have different affinities for two materials, a moving fluid (the "mobile phase") and a porous solid (the stationary phase). In FPLC the mobile phase is an aqueous solution, or "buffer". The buffer flow rate is controlled by a positive-displacement pump and is normally kept constant, while the composition of the buffer can be varied by drawing fluids in different proportions from two or more external reservoirs. The stationary phase is a resin composed of beads, usually of cross-linked agarose, packed into a cylindrical glass or plastic column. FPLC resins are available in a wide range of bead sizes and surface ligands depending on the application.

Countercurrent Chromatography

Countercurrent chromatography (CCC) is a type of liquid-liquid chromatography, where both the stationary and mobile phases are liquids and the liquid stationary phase is held stagnant by a strong centrifugal force.

Hydrodynamic Countercurrent Chromatography (CCC)

The operating principle of CCC instrument requires a column consisting of an open tube coiled around a bobbin. The bobbin is rotated in a double-axis gyratory motion (a cardioid), which causes a variable gravity (G) field to act on the column during each rotation. This motion causes the column to see one partitioning step per revolution and components of the sample separate in the column due to their partitioning coefficient between the two immiscible liquid phases used. There are many types of CCC available today. These include HSCCC (High Speed CCC) and HPCCC (High Performance CCC). HPCCC is the latest and best-performing version of the instrumentation available currently.

Centrifugal Partition Chromatography (CPC)

In the CPC (centrifugal partition chromatography or hydrostatic countercurrent chromatography) instrument, the column consists of a series of cells interconnected by ducts attached to a rotor.

This rotor rotates on its central axis creating the centrifugal field necessary to hold the stationary phase in place. The separation process in CPC is governed solely by the partitioning of solutes between the stationary and mobile phases, which mechanism can be easily described using the partition coefficients (K_D) of solutes. CPC instruments are commercially available for laboratory, pilot, and industrial-scale separations with different sizes of columns ranging from some 10 milliliters to 10 liters in volume.

Periodic Counter-Current Chromatography

In contrast to Counter current chromatography, periodic counter-current chromatography (PCC) uses a solid stationary phase and only a liquid mobile phase. It thus is much more similar to conventional affinity chromatography than to counter current chromatography. PCC uses multiple columns, which during the loading phase are connected in line. This mode allows for overloading the first column in this series without losing product, which already breaks through the column before the resin is fully saturated.

The breakthrough product is captured on the subsequent column(s). In a next step the columns are disconnected from one another. The first column is washed and eluted, while the other column(s) are still being loaded. Once the (initially) first column is re-equilibrated, it is re-introduced to the loading stream, but as last column. The process then continues in a cyclic fashion.

Chiral Chromatography

Chiral chromatography involves the separation of stereoisomers. In the case of enantiomers, these have no chemical or physical differences apart from being three-dimensional mirror images.

To enable chiral separations to take place, either the mobile phase or the stationary phase must themselves be made chiral, giving differing affinities between the analytes. Chiral chromatography HPLC columns (with a chiral stationary phase) in both normal and reversed phase are commercially available. Conventional chromatography are incapable of separating racemic mixtures of enantiomers. However, in some cases *nonracemic* mixtures of enantiomers may be separated unexpectedly by conventional liquid chromatography (e.g. HPLC without chiral mobile phase or stationary phase).

Aqueous Normal-Phase Chromatography

Aqueous normal-phase (ANP) chromatography is characterized by the elution behavior of classical normal phase mode (i.e. where the mobile phase is significantly less polar than the stationary phase) in which water is one of the mobile phase solvent system components. It is distinguished from hydrophilic interaction liquid chromatography (HILIC) in that the retention mechanism is due to adsorption rather than partitioning.

6.1.9 Applications

Chromatography is used in many fields including the pharmaceutical industry, the food and beverage industry, the chemical industry, forensic science, environment analysis, and hospitals.

6.2 PRINCIPLES OF CHROMATOGRAPHY

Chromatography is based on the principle of separating substances by distributing them between two phases: a stationary phase and a mobile phase. The stationary phase remains fixed, while the mobile phase moves through or along the stationary phase, carrying the mixture of substances with it. Each component in the mixture interacts differently with the stationary and mobile phases, depending on its chemical properties, causing the substances to travel at different rates. This difference in movement allows for the separation of the individual components of the mixture. The efficiency of separation in chromatography depends on several factors, including the polarity, solubility, and molecular size of the substances being analyzed, as well as the properties of the stationary and mobile phases. Once the components are separated, they can be detected and analyzed using various techniques, such as mass spectrometry or spectroscopy.

6.3 TYPES OF CHROMATOGRAPHY USED IN FORENSIC CHEMISTRY

There are several types of chromatography, each suited to different types of samples and forensic applications. The most commonly used forms of chromatography in forensic chemistry include gas chromatography (GC), liquid chromatography (LC), thin-layer chromatography (TLC), and high-performance liquid chromatography (HPLC). Each of these techniques has unique advantages and is used to analyze specific types of evidence.

6.3.1 Gas Chromatography (GC)

Gas chromatography is one of the most frequently used chromatographic techniques in forensic laboratories. It is particularly well-suited for the analysis of volatile and semi-volatile organic compounds, such as drugs, toxins, and explosives. In GC, the mobile phase is an inert gas (such as helium or nitrogen), while the stationary phase is a thin layer of liquid or polymer coated on the inner walls of a column.

The sample is vaporized and injected into the column, where its components are separated based on their boiling points and affinities for the stationary phase.

GC is highly efficient for separating complex mixtures and can be combined with detectors such as mass spectrometry (GC-MS) to identify the chemical composition of the separated components with great precision. For example, GC is often used in forensic toxicology to analyze blood and urine samples for the presence of drugs or alcohol. It is also used to analyze fire debris for traces of accelerants, helping investigators determine whether arson was involved.

6.3.2 Liquid Chromatography (LC)

Liquid chromatography is another widely used technique in forensic chemistry, particularly for analyzing non-volatile and thermally unstable compounds. In LC, the mobile phase is a liquid solvent, and the stationary phase is usually a solid material packed into a column. The sample is dissolved in the liquid phase and pumped through the column, where its components are separated based on their interactions with the stationary phase.

Liquid chromatography is highly versatile and can be used to analyze a wide range of substances, from drugs and toxins to explosives and inks. It is particularly useful for the analysis of biological samples, such as blood and tissue, where volatile compounds may not be present. LC can also be coupled with mass spectrometry (LC-MS) or ultraviolet-visible (UV-Vis) detection methods to provide detailed information about the chemical composition of the separated components.

6.3.3 Thin-Layer Chromatography (TLC)

Thin-layer chromatography is a simple, low-cost technique that is often used as a preliminary screening tool in forensic investigations. In TLC, the stationary phase is a thin layer of a solid material, such as silica gel or alumina, coated on a glass or plastic plate. The mobile phase is a liquid solvent that moves up the plate by capillary action, carrying the sample with it. The components of the sample separate as they travel at different rates along the stationary phase, forming distinct spots on the plate.

TLC is particularly useful for screening drugs, inks, and other organic compounds. Although it does not provide the same level of resolution as other chromatographic techniques, it is a quick and cost-effective method for identifying the presence of specific compounds in a mixture. For example, TLC is commonly used in drug testing to screen for the presence of illicit substances in blood or urine samples.

6.3.4 High-Performance Liquid Chromatography (HPLC)

High-performance liquid chromatography (HPLC) is a more advanced form of liquid chromatography that offers higher resolution, sensitivity, and speed. In HPLC, the mobile phase is pumped through the column at high pressure, allowing for faster and more efficient separation of the sample components. The stationary phase is typically a fine particle material packed into the column, providing a large surface area for interactions with the sample.

HPLC is widely used in forensic chemistry for the analysis of complex mixtures, such as drugs, toxins, and environmental contaminants. It is particularly useful for analyzing substances that are thermally unstable or non-volatile, making it an ideal technique for the analysis of biological samples. HPLC can be combined with various detectors, such as UV-Vis, fluorescence, or mass spectrometry, to provide detailed information about the separated components.

6.4 APPLICATIONS OF CHROMATOGRAPHY IN FORENSIC CHEMISTRY

Chromatography plays a crucial role in forensic chemistry, enabling the analysis of a wide variety of evidence types. Its ability to separate and identify substances in complex mixtures makes it an indispensable tool for forensic investigators. Below are some of the key applications of chromatography in forensic science.

6.4.1 Drug Analysis

One of the most common applications of chromatography in forensic chemistry is the analysis of drugs. Both illicit and prescription drugs can be present in biological samples, such as blood, urine, and hair, or in seized materials from crime scenes. Chromatography is used to identify and quantify these drugs, helping forensic scientists determine whether a suspect was under the influence of drugs or whether a particular substance is illegal.

Gas chromatography and liquid chromatography are the most commonly used techniques for drug analysis. For example, GC-MS is frequently used in forensic toxicology to analyze blood or urine samples for the presence of drugs, while HPLC is used to identify and quantify drugs in complex mixtures, such as pharmaceutical formulations.

6.4.2 Toxicology

In forensic toxicology, chromatography is used to detect and quantify poisons, toxins, and other harmful substances in biological samples. These substances may include heavy metals, pesticides, drugs, or other toxic compounds that could have contributed to a person's death or injury. Chromatography allows forensic toxicologists to identify these substances with great accuracy, providing crucial evidence in cases of poisoning or overdose.

For example, HPLC is commonly used to analyze blood and tissue samples for the presence of toxic compounds, while GC-MS is used to detect volatile substances, such as alcohol or carbon monoxide, in cases of suspected poisoning. Chromatography can also be used to analyze environmental samples, such as soil or water, for the presence of hazardous chemicals in cases of environmental contamination.

6.4.3 Explosives and Arson Investigation

Chromatography is also used in the analysis of explosives and fire debris in forensic investigations. Explosives are often composed of complex mixtures of chemicals, and chromatography allows forensic scientists to separate and identify the individual components of these mixtures. GC is particularly useful for analyzing volatile and semi-volatile compounds in explosives, while HPLC is used to analyze non-volatile components. In arson investigations, chromatography is used to analyze fire debris for traces of accelerants, such as gasoline or kerosene, that may have been used to start the fire. GC-MS is the most commonly used technique for this purpose, as it can detect and identify even trace amounts of accelerants in complex mixtures of fire debris.

6.4.4 Ink and Document Analysis

In forensic document examination, chromatography is used to analyze inks and dyes in questioned documents. TLC is often used as a preliminary screening tool to separate the different components of ink samples, allowing forensic document examiners to compare the ink used in a questioned document to a known sample. Chromatography can also be used to analyze the composition of paper and other materials used in documents, helping investigators determine whether a document is authentic or forged.

6.4.5 Fiber and Paint Analysis

Chromatography is also used to analyze trace evidence, such as fibers and paints, that may be found at crime scenes. These materials are often composed of complex mixtures of chemicals, and chromatography allows forensic scientists to separate and identify the individual components. For example, HPLC can be used to analyze the dyes in fibers, helping investigators link a fiber sample to a specific fabric or garment. Similarly, chromatography can be used to analyze the pigments and binders in paint samples, providing valuable evidence in hit-and-run cases or other crimes involving vehicles.

6.5 ADVANTAGES OF CHROMATOGRAPHY IN FORENSIC CHEMISTRY

Chromatography offers several advantages in forensic chemistry, making it one of the most widely used analytical techniques in the field. One of its primary strengths is its ability to separate complex mixtures into their individual components, allowing forensic scientists to identify and quantify substances that may be present at trace levels. This is particularly important in forensic investigations, where the evidence may be scarce or contaminated.

Another advantage of chromatography is its versatility. Chromatographic techniques can be used to analyze a wide variety of substances, from drugs and toxins to fibers and explosives. Chromatography can also be coupled with various detection methods, such as mass spectrometry or UV-Vis spectroscopy, to provide detailed information about the chemical composition of the separated components.

Finally, chromatography is a relatively fast and efficient technique, allowing forensic scientists to analyze large numbers of samples in a short amount of time. This is particularly important in forensic investigations, where timely results can be critical to solving crimes.

6.6 CONCLUSION

Chromatography is an indispensable tool in forensic chemistry, offering a powerful and versatile method for separating, identifying, and quantifying substances in complex mixtures. Its ability to analyze a wide range of materials, from biological samples to explosives, makes it a key technique in forensic investigations. Whether analyzing drugs, toxins, or trace evidence, chromatography provides forensic scientists with the detailed chemical information they need to solve crimes and deliver justice. As forensic science continues to evolve, chromatography will remain at the forefront of analytical techniques, helping to ensure the accuracy and reliability of forensic evidence.

CHAPTER 7

INORGANIC POISONS AND CONTAMINANTS

7.1 INORGANIC POISONS AND CONTAMINANTS: AN OVERVIEW

In forensic chemistry, the analysis of inorganic poisons and contaminants plays a crucial role in understanding toxicology and environmental hazards. Inorganic poisons, which include heavy metals, salts, and other non-organic compounds, can pose significant health risks and are often implicated in cases of poisoning, environmental contamination, and even homicides.

7.2 TYPES OF INORGANIC POISONS

Inorganic poisons can be categorized into several classes, each with distinct chemical properties and toxic effects. The most notable categories include heavy metals, metalloids, and certain salts.

7.2.1 Heavy Metals

Heavy metals are metallic elements with high atomic weights and densities that can be toxic to human health and the environment. Common heavy metals associated with poisoning include lead, mercury, arsenic, cadmium, and chromium. These metals can accumulate in biological tissues, leading to chronic toxicity and various health issues.

- **Lead:** Lead poisoning occurs primarily through exposure to lead-based paints, contaminated water, or occupational hazards. Lead interferes with several bodily functions, particularly affecting the nervous system, and can lead to cognitive deficits, behavioral problems, and developmental delays in children.
- **Mercury:** Mercury exists in several forms, including elemental mercury, inorganic mercury compounds, and organic mercury compounds (e.g., methylmercury). Exposure to mercury can result in severe neurological and kidney damage, as it bioaccumulates in fish and other organisms, posing risks to human health through consumption.
- **Arsenic:** Arsenic is a naturally occurring metalloid often found in groundwater. Chronic exposure to arsenic can lead to skin lesions,

respiratory problems, and increased risks of various cancers. It is also a common poison used in homicides due to its availability and undetectable nature in certain forms.

7.2.2 Metalloids

Metalloids, such as antimony and selenium, exhibit properties of both metals and non-metals. Antimony, for instance, can cause gastrointestinal distress, respiratory problems, and skin irritations upon exposure. While not as widely discussed as heavy metals, metalloids are significant in toxicological investigations, especially in cases involving contaminated environments or products.

7.2.3 Salts

Certain inorganic salts can also be toxic. For example, sodium nitrite and potassium cyanide are salts that can cause severe poisoning. Sodium nitrite is sometimes used as a preservative in food but can lead to methemoglobinemia, a condition that reduces the blood's ability to carry oxygen. Potassium cyanide, on the other hand, is a fast-acting poison that inhibits cellular respiration, leading to rapid organ failure and death.

7.3 SOURCES OF INORGANIC CONTAMINANTS

Inorganic poisons and contaminants can originate from various sources, contributing to both occupational and environmental exposure. Understanding these sources is crucial for forensic investigations.

7.3.1 Industrial Processes

Many heavy metals and inorganic contaminants are byproducts of industrial activities. Mining, smelting, and manufacturing processes can release heavy metals into the environment, leading to soil and water contamination. For instance, lead can leach into water supplies from corroded pipes or lead-based solder, particularly in older buildings.

7.3.2 Agricultural Practices

Agricultural practices can also introduce inorganic contaminants into the environment. The use of fertilizers and pesticides containing heavy metals or metalloids can lead to soil and water contamination. For example, cadmium is often found in fertilizers, and its accumulation in crops can pose significant health risks to consumers.

7.3.3 Household Products

Inorganic poisons can also be found in common household products. For instance, certain cleaning agents, paints, and batteries may contain heavy metals.

Additionally, lead-based paints in older homes remain a significant source of lead exposure, especially for children who may ingest lead dust or chips.

7.3.4 Environmental Sources

Natural sources of inorganic poisons include minerals and ores. For example, arsenic is commonly found in groundwater in some regions, particularly those with geological formations rich in arsenic-containing minerals. Understanding the geological context of an area is essential for assessing potential risks related to inorganic contaminants.

7.4 MECHANISMS OF ACTION

The mechanisms by which inorganic poisons exert their toxic effects are diverse and depend on the specific substance involved. Understanding these mechanisms is critical for assessing the health risks associated with exposure to these substances.

7.4.1 Enzyme Inhibition

Many inorganic poisons function by inhibiting key enzymes in the body. For instance, lead can inhibit enzymes involved in heme synthesis, leading to anemia and other hematological disorders. Similarly, mercury can inhibit enzymes involved in cellular respiration, resulting in oxidative stress and cell death.

7.4.2 Oxidative Stress

Some inorganic poisons can induce oxidative stress by generating reactive oxygen species (ROS) within cells. This process can lead to cellular damage, inflammation, and apoptosis. For example, cadmium is known to generate ROS, causing damage to lipids, proteins, and DNA, ultimately leading to various health issues, including cancer.

7.4.3 Disruption of Cellular Processes

Inorganic contaminants can also disrupt cellular processes by interfering with normal signaling pathways or cellular structures. For instance, arsenic can disrupt mitochondrial function, impairing energy production and contributing to cell death. Understanding these mechanisms is crucial for developing effective treatments for poisoning and environmental remediation strategies.

7.5 ANALYTICAL METHODS FOR DETECTION

The detection and quantification of inorganic poisons and contaminants in biological and environmental samples are essential in forensic chemistry. Various analytical methods are employed to achieve this, each with its advantages and limitations.

7.5.1 Atomic Absorption Spectroscopy (AAS)

Atomic absorption spectroscopy is a widely used technique for detecting heavy metals in various samples. AAS involves vaporizing the sample and measuring the absorption of light by the free atoms in the vapor. This method is particularly effective for analyzing metals like lead, mercury, and cadmium, providing sensitive and accurate measurements at low concentrations.

7.5.2 Inductively Coupled Plasma Mass Spectrometry (ICP-MS)

ICP-MS is a highly sensitive technique capable of detecting trace levels of multiple elements simultaneously. This method involves ionizing the sample and measuring the mass-to-charge ratio of the ions produced. ICP-MS is particularly useful for analyzing environmental samples, as it can detect a wide range of elements, including heavy metals and metalloids, at very low concentrations.

7.5.3 X-ray Fluorescence (XRF)

X-ray fluorescence is a non-destructive analytical technique that can be used to analyze solid samples for heavy metals and other inorganic contaminants. XRF involves exposing the sample to X-rays, causing the elements in the sample to emit characteristic X-ray fluorescence. This technique is particularly useful for analyzing soil, sediments, and paint samples.

7.5.4 High-Performance Liquid Chromatography (HPLC)

HPLC can be employed for the analysis of certain inorganic compounds, particularly those that can be dissolved in liquid.

When coupled with appropriate detection methods, HPLC can effectively separate and quantify inorganic contaminants in various matrices, such as water and biological samples.

7.6 SIGNIFICANCE IN FORENSIC INVESTIGATIONS

The analysis of inorganic poisons and contaminants holds significant importance in forensic investigations. Understanding the presence and concentration of these substances can provide critical insights into various cases, including homicides, poisoning, and environmental crime.

7.6.1 Homicide Investigations

In cases of suspected poisoning, forensic chemists often analyze biological samples for the presence of inorganic poisons. The detection of heavy metals or other inorganic contaminants can help establish cause of death and determine whether foul play was involved. For example, the identification of arsenic in a victim's tissues may indicate poisoning, prompting further investigation.

7.6.2 Environmental Crime

Environmental contamination due to inorganic poisons can have severe consequences for public health and ecosystems. Forensic chemists play a vital role in assessing contamination levels and identifying sources of pollutants in cases of environmental crime. For instance, analyzing soil and water samples for heavy metals can help determine the extent of contamination from industrial activities, guiding remediation efforts and legal actions.

7.6.3 Toxicology Reports

The analysis of inorganic contaminants is also essential in toxicology reports. These reports are crucial in legal proceedings, providing evidence regarding a person's exposure to toxic substances. Understanding the types and levels of inorganic poisons present in biological samples can influence legal outcomes in cases involving poisoning, occupational exposure, or environmental contamination.

7.6.4 Public Health Implications

Beyond individual cases, the analysis of inorganic poisons and contaminants has broader public health implications. Identifying sources of contamination and understanding the health risks associated with exposure to these substances can inform public health policies and regulatory measures. For example, monitoring lead levels in drinking water supplies can help prevent lead poisoning in vulnerable populations, particularly children.

7.7 CONCLUSION

Inorganic poisons and contaminants are significant concerns in forensic chemistry, given their potential health risks and environmental impact. Understanding the types, sources, mechanisms of action, and analytical methods for detecting these substances is essential for forensic investigations. As forensic science continues to evolve, the analysis of inorganic poisons will remain critical in solving crimes, protecting public health, and ensuring environmental safety. By employing advanced analytical techniques, forensic chemists can contribute to the pursuit of justice and the prevention of future harm.

CHAPTER 8
CONTROLLED SUBSTANCES

8.1 UNDERSTANDING THE CONTROLLED SUBSTANCES

A controlled substance is generally a drug or chemical whose manufacture, possession and use is regulated by a government, such as illicitly used drugs or prescription medications that are designated by law. Some treaties, notably the Single Convention on Narcotic Drugs, the Convention on Psychotropic Substances, and the United Nations Convention Against Illicit Traffic in Narcotic Drugs and Psychotropic Substances, provide internationally agreed-upon "schedules" of controlled substances, which have been incorporated into national laws; however, national laws usually significantly expand on these international conventions.

Some precursor chemicals used for the production of illegal drugs are also controlled substances in many countries, even though they may lack the pharmacological effects of the drugs themselves. Substances are classified according to schedules and consist primarily of potentially psychoactive substances and anabolic steroids. The controlled substances do not include many prescription items such as antibiotics.

8.1.1 Laws and Enforcement

In the United States, the Drug Enforcement Administration is the federal government agency responsible for suppressing illegal drug use and distribution by enforcing the Controlled Substances Act, which regulates both the drugs themselves and certain precursors. Some U.S. states have additional restrictions for substances which might or might not be regulated by the federal government. During the Obama Administration, the federal government also voluntarily suspended enforcement of federal laws restricting marijuana where people were operating in compliance with state law.

Some states in the U.S. have statutes against health care providers self-prescribing and/or administering substances listed in the Controlled Substance Act schedules. This does not forbid licensed providers from self-prescribing medications not on the schedules.

The term Controlled Drug in the United Kingdom (CD) is used for substances governed by the Misuse of Drugs Act 1971. Other national drug prohibition laws include the Controlled Drugs and Substances Act (Canada) and the Misuse of Drugs Act 1975 (New Zealand), among many others.

Within Europe controlled substance laws are legislated at the national rather than by the EU itself, with significant variation between countries in which and how chemicals are classified as controlled. Only drug precursor laws are legislated for at the European level.

8.1.2 Use in Research

A common misunderstanding amongst researchers is that most national laws allow the use of small amounts of a controlled substance for non-clinical / non-*in vivo* research without licences. A typical use case might be having a few milligrams or microlitres of a controlled substance within larger chemical collections (often 10K's of chemicals) for *in vitro* screening. Researchers often believe that there is some form of "research exemption" for such small amounts. This incorrect view may be further re-enforced by R&D chemical suppliers often stating and asking scientists to confirm that anything bought is for research use only.

A further misconception is that controlled substances laws simply list a few hundred substances (e.g. MDMA, Fentanyl, Amphetamine, etc.) and compliance can be achieved via checking a CAS number, chemical name or similar identifier. However, the reality is that most countries enact "generic statement" or "chemical space" laws, which aim to control all chemicals similar to the "named" substance. These either provide detailed descriptions similar to Markushes, or simply state analogues are also controlled. In addition, control of most named substances is extended to control of all of their ethers, esters, salts and stereoisomers.

Due to this complexity in legislation the identification of controlled chemicals in research is often carried out computationally, either by in house systems maintained a company's sample logistics department or by the use commercial software solutions. Automated systems are often required as many research operations can often have chemical collections running into 10Ks of molecules at the 1–5 mg scale, which are likely to include controlled substances, especially within medicinal chemistry research. These may not have been controlled when created, but they have subsequently been declared controlled.

8.1.3 Controlled Substances Act

The Controlled Substances Act (CSA) is the statute establishing federal U.S. drug policy under which the manufacture, importation, possession, use, and distribution of certain substances is regulated. It was passed by the 91st United States Congress as Title II of the Comprehensive Drug Abuse Prevention and Control Act of 1970 and signed into law by President Richard Nixon. The Act also served as the national implementing legislation for the Single Convention on Narcotic Drugs.

The legislation created five schedules (classifications), with varying qualifications for a substance to be included in each. Two federal agencies, the Drug Enforcement

Administration (DEA) and the Food and Drug Administration (FDA), determine which substances are added to or removed from the various schedules, although the statute passed by Congress created the initial listing. Congress has sometimes scheduled other substances through legislation such as the Hillory J. Farias and Samantha Reid Date-Rape Prevention Act of 2000, which placed gamma hydroxybutyrate (GHB) in Schedule I and sodium oxybate (the isolated sodium salt in GHB) in Schedule III when used under an FDA New Drug Application (NDA) or Investigational New Drug (IND). Classification decisions are required to be made on criteria including potential for abuse (an undefined term), currently accepted medical use in treatment in the United States, and international treaties.

History

The nation first outlawed addictive drugs in the early 1900s and the International Opium Convention helped lead international agreements regulating trade. The Pure Food and Drug Act (1906) was the beginning of over 200 laws concerning public health and consumer protections. Others were the Federal Food, Drug, and Cosmetic Act (1938), and the Kefauver Harris Amendment of 1962.

In 1969, President Richard Nixon announced that the Attorney General, John N. Mitchell, was preparing a comprehensive new measure to more effectively meet the narcotic and dangerous drug problems at the federal level by combining all existing federal laws into a single new statute. With the help of White House Counsel head, John Dean; the executive director of the Shafer Commission, Michael Sonnenreich; and the Director of the BNDD, John Ingersoll creating and writing the legislation, Mitchell was able to present Nixon with the bill.

The CSA not only combined existing federal drug laws and expanded their scope, but it also changed the nature of federal drug law policies and expanded federal law enforcement pertaining to controlled substances. Title II, Part F of the Comprehensive Drug Abuse Prevention and Control Act of 1970 established the National Commission on Marijuana and Drug Abuse—known as the Shafer Commission after its chairman, Raymond P. Shafer—to study cannabis abuse in the United States. During his presentation of the commission's First Report to Congress, Sonnenreich and Shafer recommended the decriminalization of marijuana in small amounts, with Shafer stating,

- [T]he criminal law is too harsh a tool to apply to personal possession even in the effort to discourage use. It implies an overwhelming indictment of the behavior which we believe is not appropriate. The actual and potential harm of use of the drug is not great enough to justify intrusion by the criminal law into private behavior, a step which our society takes only with the greatest reluctance.

Rufus King notes that this stratagem was similar to that used by Harry Anslinger when he consolidated the previous anti-drug treaties into the Single Convention and took the opportunity to add new provisions that otherwise might have been unpalatable to the international community. According to David T. Courtwright, "the Act was part of an omnibus reform package designed to rationalize, and in some respects to liberalize, American drug policy." (Courtwright noted that the Act became, not libertarian, but instead repressionistic to the point of tyrannical in its intent; a cruel and/or arbitrary exercise of power). It eliminated mandatory minimum sentences and provided support for drug treatment and research.

King notes that the rehabilitation clauses were added as a compromise to Senator Jim Hughes, who favored a moderate approach. The bill, as introduced by Senator Everett Dirksen, ran to 91 pages. While it was being drafted, the Uniform Controlled Substances Act, to be passed by state legislatures, was also being drafted by the Department of Justice; its wording closely mirrored the Controlled Substances Act.

Amendments, 1970–2018

Since its enactment in 1970, the Act has been amended numerous times:

- The 1976 Medical Device Regulation Act.
- The Psychotropic Substances Act of 1978 added provisions implementing the Convention on Psychotropic Substances.
- The Controlled Substances Penalties Amendments Act of 1984.
- The 1986 Federal Analog Act for chemicals "substantially similar" in Schedule I and II to be listed.
- The 1988 Chemical Diversion and Trafficking Act (implemented August 1, 1989 as Article 12) added provisions implementing the United Nations Convention Against Illicit Traffic in Narcotic Drugs and Psychotropic Substances that went into force on November 11, 1990.
- 1990 The Anabolic Steroids Act, passed as part of the Crime Control Act of 1990, which placed anabolic steroids into Schedule III.
- The 1993 Domestic Chemical Diversion and Control Act (effective on April 16, 1994) in response to methamphetamine trafficking.
- The Hillory J. Farias and Samantha Reid Date-Rape Prevention Act of 2000 placed gamma hydroxybutyrate (GHB) in Schedule I and sodium oxybate (the isolated sodium salt in GHB) in Schedule III when used under an FDA NDA or IND.
- The 2008 Ryan Haight Online Pharmacy Consumer Protection Act.
- The 2010 Electronic Prescriptions for Controlled Substances (EPCS).

- The 2012 Synthetic Drug Abuse Prevention Act Subtitle D - synthetic drugs, added several Markush like statements that describes synthetic cannabinoid chemical space that are also controlled as Schedule 1 substances. However, since then many new synthetic cannabinoids not covered by this act have emerged.
- The 2010 Secure and Responsible Drug Disposal Act (effective on October 12, 2010), to allow pharmacies to operate take-back programs for controlled substance medications in response to the US opioid epidemic.
- The 2017 Protecting Patient Access to Emergency Medications Act (PPAEMA) amended Section 33 of the CSA to include DEA registration for Emergency Medical Service (EMS) agencies, approved uses of standing orders, and requirements for the maintenance and administration of controlled substances used by EMS agencies.
- In 2018 the act was also amended to describe and control all chemical space related to Fentanyl like chemicals using Markush like notation, the first time Markush like statement were directly used in the act itself.

Statute Content

The Controlled Substances Act consists of two subchapters. Subchapter I defines Schedules I–V, lists chemicals used in the manufacture of controlled substances, and differentiates lawful and unlawful manufacturing, distribution, and possession of controlled substances, including possession of Schedule I drugs for personal use; this subchapter also specifies the dollar amounts of fines and durations of prison terms for violations.

Subchapter II describes the laws for exportation and importation of controlled substances, again specifying fines and prison terms for violations.

Enforcement Authority

The Drug Enforcement Administration was established in 1973, combining the Bureau of Narcotics and Dangerous Drugs (BNDD) and Customs' drug agents.

Proceedings to add, delete, or change the schedule of a drug or other substance may be initiated by the DEA, the Department of Health and Human Services (HHS), or by petition from any interested party, including the manufacturer of a drug, a medical society or association, a pharmacy association, a public interest group concerned with drug abuse, a state or local government agency, or an individual citizen. When a petition is received by the DEA, the agency begins its own investigation of the drug. The DEA may begin an investigation of a drug at any time based upon information received from laboratories, state and local law enforcement and regulatory agencies, or other sources of information. Once the DEA has collected the necessary data, the Deputy Administrator of DEA, requests from HHS a scientific and medical evaluation and recommendation as to whether the drug or other substance should be controlled or removed from control.

This request is sent to the Assistant Secretary of Health of HHS. Then, HHS solicits information from the Commissioner of the Food and Drug Administration and evaluations and recommendations from the National Institute on Drug Abuse and, on occasion, from the scientific and medical community at large. The Assistant Secretary, by authority of the Secretary, compiles the information and transmits back to the DEA a medical and scientific evaluation regarding the drug or other substance, a recommendation as to whether the drug should be controlled, and in what schedule it should be placed.

The HHS recommendation on scheduling is binding to the extent that if HHS recommends, based on its medical and scientific evaluation, that the substance not be controlled, then the DEA may not control the substance. Once the DEA has received the scientific and medical evaluation from HHS, the DEA Administrator evaluates all available data and makes a final decision whether to propose that a drug or other substance be controlled and into which schedule it should be placed. Under certain circumstances, the Government may temporarily schedule a drug without following the normal procedure.

An example is when international treaties require control of a substance. 21 U.S.C. § 811(h) allows the Attorney General to temporarily place a substance in Schedule I "to avoid an imminent hazard to the public safety". Thirty days' notice is required before the order can be issued, and the scheduling expires after a year. The period may be extended six months if rulemaking proceedings to permanently schedule the drug are in progress. In any case, once these proceedings are complete, the temporary order is automatically vacated. Unlike ordinary scheduling proceedings, such temporary orders are not subject to judicial review.

The CSA creates a closed system of distribution for those authorized to handle controlled substances. The cornerstone of this system is the registration of all those authorized by the DEA to handle controlled substances. All individuals and firms that are registered are required to maintain complete and accurate inventories and records of all transactions involving controlled substances, as well as security for the storage of controlled substances.

Treaty Obligations

The Congressional findings in 21 USC §§ 7)801), 801a(2), and 801a(3) state that a major purpose of the CSA is to "enable the United States to meet all of its obligations" under international treaties. The CSA bears many resemblances to these Conventions. Both the CSA and the treaties set out a system for classifying controlled substances in several schedules in accordance with the binding scientific and medical findings of a public health authority. Under 21 U.S.C. § 811 of the CSA, that authority is the Secretary of Health and Human Services (HHS). Under Article 3 of the Single Convention and Article 2 of the Convention on Psychotropic Substances, the World Health Organization is that authority.

The domestic and international legal nature of these treaty obligations must be considered in light of the supremacy of the United States Constitution over treaties or acts and the equality of treaties and Congressional acts. In *Reid v. Covert* the Supreme Court of the United States addressed both these issues directly and clearly holding:

- [N]o agreement with a foreign nation can confer power on the Congress, or on any other branch of Government, which is free from the restraints of the Constitution.
- Article VI, the Supremacy Clause of the Constitution, declares:
  - "This Constitution, and the Laws of the United States which shall be made in Pursuance thereof, and all Treaties made, or which shall be made, under the Authority of the United States, shall be the supreme Law of the Land;..."

There is nothing in this language which intimates that treaties and laws enacted pursuant to them do not have to comply with the provisions of the Constitution. Nor is there anything in the debates which accompanied the drafting and ratification of the Constitution which even suggests such a result. These debates, as well as the history that surrounds the adoption of the treaty provision in Article VI, make it clear that the reason treaties were not limited to those made in "pursuance" of the Constitution was so that agreements made by the United States under the Articles of Confederation, including the important peace treaties which concluded the Revolutionary War, would remain in effect. It would be manifestly contrary to the objectives of those who created the Constitution, as well as those who were responsible for the Bill of Rights—let alone alien to our entire constitutional history and tradition—to construe Article VI as permitting the United States to exercise power under an international agreement without observing constitutional prohibitions. In effect, such construction would permit amendment of that document in a manner not sanctioned by Article V. The prohibitions of the Constitution were designed to apply to all branches of the National Government, and they cannot be nullified by the Executive or by the Executive and the Senate combined.

There is nothing new or unique about what we say here. This Court has regularly and uniformly recognized the supremacy of the Constitution over a treaty. For example, in *Geofroy v. Riggs*, 133 U. S. 258, 133 U. S. 267, it declared:

- "The treaty power, as expressed in the Constitution, is in terms unlimited except by those restraints which are found in that instrument against the action of the government or of its departments, and those arising from the nature of the government itself and of that of the States. It would not be contended that it extends so far as to authorize what the Constitution forbids, or a change in the character of the government, or in that of one of the States, or a cession of any portion of the territory of the latter, without its consent."
- This Court has repeatedly taken the position that an Act of Congress, which must comply with the Constitution, is on a full parity with a treaty, and that, when a statute which is subsequent in time is inconsistent with a treaty, the statute to the extent of conflict renders the treaty null. It would be completely anomalous to say that a treaty need not comply with the Constitution when such an agreement can be overridden by a statute that must conform to that instrument.

According to the Cato Institute, these treaties only bind (legally obligate) the United States to comply with them as long as that nation agrees to remain a state party to these treaties. The U.S. Congress and the President of the United States have the absolute sovereign right to withdraw from or abrogate at any time these two instruments, in accordance with said nation's Constitution, at which point these treaties will cease to bind that nation in any way, shape, or form.

A provision for automatic compliance with treaty obligations is found at 21 U.S.C. § 811(d), which also establishes mechanisms for amending international drug control regulations to correspond with HHS findings on scientific and medical issues.

If control of a substance is mandated by the Single Convention, the Attorney General is required to "issue an order controlling such drug under the schedule he deems most appropriate to carry out such obligations," without regard to the normal scheduling procedure or the findings of the HHS Secretary. However, the Secretary has great influence over any drug scheduling proposal under the Single Convention, because 21 U.S.C. § 811(d)(2)(B) requires the Secretary the power to "evaluate the proposal and furnish a recommendation to the Secretary of State which shall be binding on the representative of the United States in discussions and negotiations relating to the proposal."

Similarly, if the United Nations Commission on Narcotic Drugs adds or transfers a substance to a schedule established by the Convention on Psychotropic Substances, so that current U.S. regulations on the drug do not meet the treaty's requirements, the Secretary is required to issue a recommendation on how the substance should be scheduled under the CSA. If the Secretary agrees with the commission's scheduling decision, he can recommend that the Attorney General initiate proceedings to reschedule the drug accordingly.

If the HHS Secretary disagrees with the UN controls, the Attorney General must temporarily place the drug in Schedule IV or V (whichever meets the minimum requirements of the treaty) and exclude the substance from any regulations not mandated by the treaty. The Secretary is required to request that the Secretary of State take action, through the commission or the UN Economic and Social Council, to remove the drug from international control or transfer it to a different schedule under the convention. The temporary scheduling expires as soon as control is no longer needed to meet international treaty obligations.

This provision was invoked in 1984 to place Rohypnol (flunitrazepam) in Schedule IV. The drug did not then meet the Controlled Substances Act's criteria for scheduling; however, control was required by the Convention on Psychotropic Substances. In 1999, an FDA official explained to Congress:

- Rohypnol is not approved or available for medical use in the United States, but it is temporarily controlled in Schedule IV pursuant to a treaty obligation under the 1971 Convention on Psychotropic Substances. At the time flunitrazepam was placed temporarily in Schedule IV (November 5, 1984), there was no evidence of abuse or trafficking of the drug in the United States.

The Cato Institute's *Handbook for Congress* calls for repealing the CSA, an action that would likely bring the United States into conflict with international law, were the United States not to exercise its sovereign right to withdraw from and/or abrogate the Single Convention on Narcotic Drugs and/or the 1971 Convention on Psychotropic Substances prior to repealing the Controlled Substances Act. The exception would be if the U.S. were to claim that the treaty obligations violate the United States Constitution. Many articles in these treaties—such as Article 35 and Article 36 of the Single Convention—are prefaced with phrases such as "Having due regard to their constitutional, legal and administrative systems, the Parties shall..." or "Subject to its constitutional limitations, each Party shall..." According to former United Nations Drug Control Programme Chief of Demand Reduction Cindy Fazey, "This has been used by the USA not to implement part of article 3 of the 1988 Convention, which prevents inciting others to use narcotic or psychotropic drugs, on the basis that this would be in contravention of their constitutional amendment guaranteeing freedom of speech".

Schedules of Controlled Substances

There are five different schedules of controlled substances, numbered I–V. The CSA describes the different schedules based on three factors:

- **Potential for Abuse:** How likely is this drug to be abused?
- **Accepted Medical Use:** Is this drug used as a treatment in the United States?
- **Safety and Potential for Addiction:** Is this drug safe? How likely is this drug to cause addiction? What kinds of addiction?

Schedule I

Schedule I substances are described as those that have all of the following findings:

- The drug or other substance has a high potential for abuse.
- The drug or other substance has no currently accepted medical use in treatment in the United States.
- There is a lack of accepted safety for use of the drug or other substance under medical supervision.

No prescriptions may be written for Schedule I substances, and such substances are subject to production quotas which the DEA imposes. Under the DEA's interpretation of the CSA, a drug does not necessarily have to have the same "high potential for abuse" as heroin, for example, to merit placement in Schedule I:

- [W]hen it comes to a drug that is currently listed in schedule I, if it is undisputed that such drug has no currently accepted medical use in treatment in the United States and a lack of accepted safety for use under medical supervision, and it is further undisputed that the drug has *at least some potential for abuse sufficient to warrant control under the CSA*, the drug must remain in schedule I. In such circumstances, placement of the drug in schedules II through V would conflict with the CSA since such drug would not meet the criterion of "a currently accepted medical use in treatment in the United States." 21 USC 812(b). (emphasis added)—*Drug Enforcement Administration, Notice of denial of petition to reschedule marijuana (2001)*

Drugs listed in this control schedule include:

- αMT (alpha-methyltryptamine), a psychedelic, stimulant, and entactogen drug of the tryptamine class that was originally developed as an antidepressant by workers at Upjohn in the 1960s.
- BZP (benzylpiperazine), a synthetic stimulant once sold as a designer drug. It has been shown to be associated with an increase in seizures if taken alone. Although the effects of BZP are not as potent as MDMA, it can produce neuroadaptations that can cause an increase in the potential for abuse of this drug.

- Cathinone, an amphetamine-like stimulant found in the shrub *Catha edulis* (khat).
- DMT (dimethyltryptamine), a naturally occurring psychedelic drug that is widespread throughout the plant kingdom and endogenous to the human body. DMT is the main psychoactive constituent in the psychedelic South American brew, ayahuasca, for which the UDV are granted exemption from DMT's schedule I status on the grounds of religious freedom.
- Etorphine, a semi-synthetic opioid possessing an analgesic potency approximately 1,000–3,000 times that of morphine.
- GHB (gamma-Hydroxybutyric acid), a general anesthetic and treatment for narcolepsy-cataplexy and alcohol withdrawal with a limited safe dosage range and poor ability to control pain when used as an anesthetic (severely limiting its usefulness). It was placed in Schedule I in March 2000 after widespread recreational use led to increased emergency room visits, hospitalizations, and deaths. A specific formulation of this drug is also listed in Schedule III for limited uses, under the trademark Xyrem.
- Heroin is the brand name for *diacetylmorphine* or *morphine diacetate*, which is an inactive prodrug that exerts its effects after being converted into the major active metabolite morphine, and the minor metabolite 6-MAM - which itself is also rapidly converted to morphine. Some European countries still use it as a potent pain reliever in terminal cancer patients, and as second option, after morphine sulfate; it is about twice as potent, by weight, as morphine and, indeed, becomes morphine upon injection into the bloodstream. The two acetyl groups attached to the morphine make a prodrug which delivers morphine to the opioid receptors twice as fast as morphine can.
- Ibogaine, a naturally occurring psychoactive substance found in plants in the family Apocynaceae. Some countries in North America use ibogaine as an alternative medicine treatment for opioid drug addiction. Ibogaine is also used for medicinal and ritual purposes within African spiritual traditions of the Bwiti.
- LSD (lysergic acid diethylamide), a semi-synthetic psychedelic drug famous for its involvement in the counterculture of the 1960s.
- Marijuana and its cannabinoids. Pure (–)-trans-Δ9-tetrahydrocannabinol is also listed in Schedule III for limited uses, under the trademark Marinol. As a result of ballot initiatives, many states have made recreational and

medical use of marijuana legal, while other states have decriminalized possession of small amounts. Such measures operate only on state laws, and have no effect on federal law. Whether such users would actually be prosecuted under federal law is a separate question with no definitive answer. Given the widespread medicinal use of cannabis, the maintenance of its Schedule I classification has been controversial, with many calling for a reclassification or holistic federal decriminalization. As of April 30, 2024, cannabis was set to be reclassified by the DEA as a Schedule III controlled substance.

- MDMA ("ecstasy" or "molly"), a stimulant, psychedelic, and entactogenic drug which initially garnered attention in psychedelic therapy as a treatment for post-traumatic stress disorder (PTSD). The medical community originally agreed upon placing it as a Schedule III substance, but the government denied this suggestion, despite two court rulings by the DEA's administrative law judge that placing MDMA in Schedule I was illegal. It was temporarily unscheduled after the first administrative hearing from December 22, 1987 – July 1, 1988.
- Mescaline, a naturally occurring psychedelic drug and the main psychoactive constituent of peyote (*Lophophora williamsii*), San Pedro cactus (*Echinopsis pachanoi*), and Peruvian torch cactus (*Echinopsis peruviana*).
- Methaqualone (Quaalude, Sopor, Mandrax), a sedative that was previously used for similar purposes as barbiturates, until it was rescheduled.
- Peyote (*Lophophora williamsii*), a cactus growing in nature primarily in northeastern Mexico; one of the few plants specifically scheduled, with a narrow exception to its legal status for religious use in Native American churches.
- Psilocybin and psilocin, naturally occurring psychedelic drugs and the main psychoactive constituents of psilocybin mushrooms.
- Controlled substance analogues intended for human consumption, as defined by the Federal Analogue Act.

In addition to the named substance, usually all possible ethers, esters, salts and stereoisomers of these substances are also controlled and also 'analogues', which are chemically similar chemicals.

Schedule II

Schedule II substances are those that have the following findings:

- The drug or other substance has a high potential for abuse
- The drug or other substance has a currently accepted medical use in treatment in the United States, or a currently accepted medical use with severe restrictions
- Abuse of the drug or other substances may lead to severe psychological or physical dependence.

Except when dispensed directly to an ultimate user by a practitioner other than a pharmacist, no controlled substance in Schedule II, which is a prescription drug as determined under the Federal Food, Drug, and Cosmetic Act (21 USC 301 *et seq.*), may be dispensed without the written or electronically transmitted (21 CFR 1306.08) prescription of a practitioner, except that in emergency situations, as prescribed by the Secretary by regulation after consultation with the Attorney General, such drug may be dispensed upon oral prescription in accordance with section 503(b) of that Act (21 USC 353 (b)). With exceptions, an original prescription is always required even though faxing in a prescription in advance to a pharmacy by a prescriber is allowed. Prescriptions shall be retained in conformity with the requirements of section 827 of this title. No prescription for a controlled substance in Schedule II may be refilled.

These drugs vary in potency: for example fentanyl is about 80 times as potent as morphine (heroin is roughly two times as potent). More significantly, they vary in nature. Pharmacology and CSA scheduling have a weak relationship.

Because refills of prescriptions for Schedule II substances are not allowed, it can be burdensome to both the practitioner and the patient if the substances are to be used on a long-term basis. To provide relief, in 2007, 21 CFR 1306.12 was amended (at 72 FR 64921) to allow practitioners to write up to three prescriptions at once, to provide up to a 90-day supply, specifying on each the earliest date on which it may be filled.

Drugs in this schedule include:

- Amphetamine drugs including Adderall, Dextroamphetamine (Dexedrine), Lisdexamfetamine (Vyvanse): treatment of ADHD, narcolepsy, severe obesity (limited use, dextroamphetamine only), binge eating disorder (lisdexamfetamine only). Originally placed in Schedule III, but moved to Schedule II in 1978 as part of the Psychotropic Substances Act.
- Barbiturates (short-acting), such as pentobarbital.
- Cocaine: used as a topical anesthetic or local anesthetic and to stop severe epistaxis.

- Codeine (pure) and any drug for non-parenteral administration containing the equivalent of more than 90 mg of codeine per dosage unit.
- Diphenoxylate (pure).
- Fentanyl and most other strong pure opioid agonists, e.g. levorphanol.
- Hydrocodone in *any* formulation since October 2014 (examples include Vicodin, Norco, Tussionex). Prior to October 2014, formulations containing hydrocodone and over-the-counter analgesics such as Acetaminophen and Ibuprofen were Schedule III.
- Hydromorphone (semi-synthetic opioid; active ingredient in Dilaudid, Palladone).
- Methadone: treatment of heroin addiction, extreme chronic pain.
- Methamphetamine: treatment of ADHD (rare), severe obesity (limited use) under the brandname Desoxyn.
- Methylphenidate (Ritalin, Concerta), Dexmethylphenidate (Focalin): treatment of ADHD, narcolepsy.
- Morphine: a pain medication of the opiate family.
- Nabilone (Cesamet) – A synthetic cannabinoid. An analogue to dronabinol (Marinol) which is a Schedule III drug.
- Opium tincture (Laudanum): a potent antidiarrheal.
- Oxycodone (semi-synthetic opioid; active ingredient in Percocet, OxyContin, and Percodan).
- Oxymorphone (semi-synthetic opioid; active ingredient in Opana).
- Nembutal (Pentobarbital) – barbiturate medication originally developed for narcolepsy; primarily used today for physician assisted suicide and euthanasia of animals.
- Pethidine (USAN: Meperidine; Demerol).
- Phencyclidine (PCP) - Formerly used as veterinary anesthetic under the trade name Sernylan and before then as an injectable anesthetic under the trade name Sernyl.
- Secobarbital (Seconal).
- Tapentadol (Nucynta) – A drug with mixed opioid agonist and norepinephrine re-uptake inhibitor activity.

Schedule III

Schedule III substances are those that have the following findings:

- The drug or other substance has a potential for abuse less than the drugs or other substances in Schedules I and II.
- The drug or other substance has a currently accepted medical use in treatment in the United States.
- Abuse of the drug or other substance may lead to moderate or low physical dependence or high psychological dependence.

Except when dispensed directly by a practitioner, other than a pharmacist, to an ultimate user, no controlled substance in Schedule III or IV, which is a prescription drug as determined under the Federal Food, Drug, and Cosmetic Act (21 USC 301 *et seq.*), may be dispensed without a written, electronically transmitted, or oral prescription in conformity with section 503(b) of that Act (21 USC 353 (b)). Such prescriptions may not be filled or refilled more than six months after the date thereof or be refilled more than five times after the date of the prescription unless renewed by the practitioner.

A prescription for controlled substances in Schedules III, IV, and V issued by a practitioner, may be communicated either orally, in writing, electronically transmitted or by facsimile to the pharmacist, and may be refilled if so authorized on the prescription or by call-in. Control of wholesale distribution is somewhat less stringent than Schedule II drugs. Provisions for emergency situations are less restrictive within the "closed system" of the Controlled Substances Act than for Schedule II though no schedule has provisions to address circumstances where the closed system is unavailable, nonfunctioning or otherwise inadequate.

Drugs in this schedule include:

- Ketamine, a drug originally developed as a safer, shorter-acting replacement for PCP (mainly for use as a human anesthetic) but has since become popular as a veterinary and pediatric anesthetic;
- Anabolic steroids (including prohormones such as androstenedione); the specific end molecule testosterone in many of its forms (Androderm, AndroGel, Testosterone Cypionate, and Testosterone Enanthate) are labeled as Schedule III while low-dose testosterone when compounded with estrogen derivatives have been exempted (from scheduling) by the FDA;
- Intermediate-acting barbiturates, such as talbutal or butalbital;
- Buprenorphine (semi-synthetic opioid; active in Suboxone, Subutex);
- Dihydrocodeine when compounded with other substances, to a certain dosage and concentration.

- FDA-approved sodium oxybate products (e.g. Xyrem, Xywav and Lumryz)—preparations of GHB used to treat narcolepsy. These products are in Schedule III but with a restricted distribution system. All other forms or preparations of GHB are in Schedule I.
- Marinol, synthetically prepared tetrahydrocannabinol (officially referred to by its INN, dronabinol) used to treat nausea and vomiting caused by chemotherapy, as well as appetite loss caused by AIDS.
- Paregoric, an antidiarrheal and anti-tussive, which contains opium combined with camphor (which makes it less addiction-prone than laudanum, which is in Schedule II).
- Phendimetrazine Tartrate, a stimulant synthesized for use as an anorexiant.
- Benzphetamine HCl (Didrex), a stimulant designed for use as an anorexiant.
- Fast-acting barbiturates such as secobarbital (Seconal) and pentobarbital (Nembutal), when combined with one or more additional active ingredient(s) not in Schedule II (e.g., Carbrital (no longer marketed), a combination of pentobarbital and carbromal).
- Ergine (lysergic acid amide), listed as a sedative but also has psychedelic effects such as visual and auditory effects. An inefficient precursor to its *N,N*-diethyl analogue, LSD, ergine occurs naturally in the seeds of the common garden flowers *Turbina corymbosa*, *Ipomoea tricolor*, and *Argyreia nervosa*.
- Perampanel (Fycompa), an anticonvulsant

Schedule IV

Placement on schedules; findings required Schedule IV substances are those that have the following findings:

- The drug or other substance has a low potential for abuse relative to the drugs or other substances in Schedule III.
- The drug or other substance has a currently accepted medical use in treatment in the United States.
- Abuse of the drug or other substance may lead to limited physical dependence or psychological dependence relative to the drugs or other substances in Schedule III.

Control measures are similar to Schedule III. Prescriptions for Schedule IV drugs may be refilled up to five times within a six-month period. A prescription for controlled substances in Schedules III, IV, and V issued by a practitioner, may be communicated either orally, in writing, electronically transmitted or by facsimile to the pharmacist, and may be refilled if so authorized on the prescription or by call-in.

Drugs in this schedule include:

- Benzodiazepines, such as alprazolam (Xanax), chlordiazepoxide (Librium), clonazepam (Klonopin), diazepam (Valium), midazolam (Versed), and Lorazepam (Ativan), as well as:
  - temazepam (Restoril) (some states require specially coded prescriptions for temazepam);
  - flunitrazepam (Rohypnol) (flunitrazepam is not FDA approved making it an illegal drug in the United States); and
  - oxazepam (Serax, Serepax, Seresta, Alepam, Opamox, Oxamin).
- The benzodiazepine-like Z-drugs: zolpidem (Ambien), zopiclone (Imovane), eszopiclone (Lunesta), and zaleplon (Sonata) (zopiclone is not commercially available in the U.S.).
- Chloral hydrate, a sedative-hypnotic.
- Long-acting barbiturates such as phenobarbital.
- Some partial agonist opioid analgesics, such as pentazocine (Talwin).
- The eugeroic drug modafinil (sold in the U.S. as Provigil) as well as its *(R)*-enantiomer armodafinil (sold in the U.S. as Nuvigil).
- Difenoxin, an antidiarrheal drug, when combined with atropine (such as Motofen) (difenoxin is 2–3 times more potent than diphenoxylate, the active ingredient in Lomotil, which is in Schedule V).
- Tramadol (Ultram), an opioid analgesic.
- Carisoprodol (Soma) has become a Schedule IV medication as of 11 January 2012.
- Suvorexant and Lemborexant, orexinergic sedatives.

Schedule V

Schedule V substances are those that have the following findings:

- The drug or other substance has a low potential for abuse relative to the drugs or other substances in schedule IV.
- The drug or other substance has a currently accepted medical use in treatment in the United States.
- Abuse of the drug or other substance may lead to limited physical dependence or psychological dependence relative to the drugs or other substances in schedule IV.

No controlled substance in Schedule V which is a drug may be distributed or dispensed other than for a medical purpose. A prescription for controlled substances in Schedules III, IV, and V issued by a practitioner, may be communicated either orally, in writing, electronically transmitted or by facsimile to the pharmacist, and may be refilled if so authorized on the prescription or by call-in.

Drugs in this schedule include:

- Cough suppressants containing small amounts of codeine (e.g., promethazine+codeine);
- Preparations containing small amounts of opium or diphenoxylate (used to treat diarrhea);
- Some anticonvulsants, such as pregabalin (Lyrica), lacosamide (Vimpat), brivaracetam (Briviact), and retigabine (ezogabine) (Potiga/Trobalt);
- Pyrovalerone (used to treat chronic fatigue and as an appetite suppressant for weight loss);
- Some centrally-acting antidiarrheals, such as diphenoxylate (Lomotil) when mixed with atropine (to make it poisonous, if taken at euphoria-inducing dosages). Difenoxin with atropine (Motofen) has been moved to Schedule IV. Without atropine, these drugs are in Schedule II; and
- Cannabidiol, only in a cannabis-derived pharmaceutical formulation marketed by GW Pharmaceuticals as Epidiolex. Other CBD formulations remain Schedule I, except for those derived from hemp which are unscheduled but still FDA-regulated.

Controlled by other Federal Laws for Legal Recreational Use

These psychoactive drugs are not controlled by the act, and are also allowed for sale intended for recreational use at the federal level (others are allowed for sale as dietary supplements, but not specifically regulated or intended for recreational use):

- Alcohol (ethanol), a sedative found in alcoholic drinks. Per the National Minimum Drinking Age Act (which is voluntarily abided by all 50 U.S. states), sale is limited to persons 21-years-old and above only. Sale regulated by the Bureau of Alcohol, Tobacco, Firearms and Explosives (ATF) and less commonly the Food and Drug Administration (FDA). Alcohol was formerly illegal under the Eighteenth Amendment to the Constitution from 1919, until the Twenty-first Amendment repealed it in 1933.
- Caffeine, a stimulant found in coffee, chocolate; and some teas and soft drinks. It is regulated by the FDA under the Federal Food, Drug,

and Cosmetic Act, and drinks cannot contain more than 200 parts per million (0.02%) of caffeine. There is no federal age restriction for caffeine-containing products. Also available medically in some pain medications (usually in combination with other drugs, like in aspirin/acetaminophen/caffeine).

- Nicotine, a stimulant found in tobacco (including cigarettes and cigars) and electronic cigarettes. Also used medically in nicotine replacement therapy. The minimum purchasing age of tobacco and e-cigarettes in the United States is 21-years-old, per the Synar amendment to the Public Health Service Act. Sales are regulated by the ATF and FDA.

Regulation of Precursors

The Controlled Substances Act also provides for federal regulation of precursors used to manufacture some of the controlled substances. The DEA list of chemicals is actually modified when the United States Attorney General determines that illegal manufacturing processes have changed.

In addition to the CSA, due to pseudoephedrine (PSE) and ephedrine being widely used in the manufacture of methamphetamine, the U.S. Congress passed the Methamphetamine Precursor Control Act which places restrictions on the sale of any medicine containing pseudoephedrine. That bill was then superseded by the Combat Methamphetamine Epidemic Act of 2005, which was passed as an amendment to the Patriot Act renewal and included wider and more comprehensive restrictions on the sale of PSE-containing products. This law requires customer signature of a "log-book" and presentation of valid photo ID in order to purchase PSE-containing products from all retailers.

Additionally, the law restricts an individual to the retail purchase of no more than three packages or 3.6 grams of such product per day per purchase – and no more than 9 grams in a single month. A violation of this statute constitutes a misdemeanor. Retailers now commonly require PSE-containing products to be sold behind the pharmacy or service counter. This affects many preparations which were previously available over-the-counter without restriction, such as Actifed and its generic equivalents.

Research Exemptions

A common misunderstanding amongst researchers is that most national laws (including the Controlled Substance Act) allows the supply/use of small amounts of a controlled substance for non-clinical / non-in vivo research without licenses. A typical use case might be having a few milligrams or microlitres of a controlled substance within larger chemical collections (often tens of thousands of chemicals) for in vitro screening or sale. Researchers often believe that there is some form of "research exemption" for such small amounts. This incorrect view may be further re-enforced by R&D chemical suppliers often stating and asking scientists to confirm that anything bought is for research use only.

A further misconception is that the Controlled Substances Act simply lists a few hundred substances (e.g. MDMA, Fentanyl, Amphetamine, etc.) and compliance can be achieved via checking a CAS number, chemical name or similar identifier. However, the reality is that in most cases all ethers, esters, salts and stereoisomers are also controlled and it is impossible to simply list all of these. The act contains several "generic statements" or "chemical space" laws, which aim to control all chemicals similar to the "named" substance, these provide detailed descriptions similar to Markushes, these include ones for Fentanyl and also synthetic cannabinoids.

Due to this complexity in legislation, the identification of controlled chemicals in research or chemical supply is often carried out computationally on the chemical structure, either by in-house systems maintained a company or by the use of commercial software solutions. Automated systems are often required as many research operations can have collections of 10,000–100,000 different substances at the 1–5 milligram scale, which are likely to include controlled substances, especially within medicinal chemistry research, even if the core focus of the company is not narcotic or psychotropic drugs. These may not have been controlled when created, but they have subsequently been declared controlled, or fall within chemical space close to known controlled substances, or are used as tool compounds, precursors or synthetic intermediates to a controlled substance.

Analogues vs. Markush Descriptions

Historically, in an attempt to prevent psychoactive chemicals which are chemically similar to controlled substance, but not specifically controlled by it, the CSA also controls "analogues" of many listed controlled substances. The definition of what 'analogue' means is kept deliberately vague, presumably to make it harder to circumvent this rule, as it's not clear what is / is not controlled, thus placing an element of risk and deterrent in those performing the supply. It is up to the courts to then decide whether a specific chemical is an analogue, often via a 'battle of experts' for the defense and prosecution which can lead to extended and more uncertain prosecutions. The use of the 'analogue' definition also make it more difficult for companies involved in the legitimate supply of chemicals for research and industrial purposes to know whether a chemical is regulated under the CSA

Starting in 2012, with the Synthetic drug abuse prevention act, and later an amendment to the CSA in 2018 defining fentanyl chemical space, the CSA started to use Markush descriptions to clearly define what analogues or chemical space is controlled. These chemical space, chemical family, generic statements or markush statements (depending on the legislation terminology) have been used for many years by other countries, notably the UK in the Misuse of Drugs Act.

These have the advantage of clearly defining what is controlled, making prosecutions easier and compliance by legitimate companies simpler. However the downside is that these tend to be harder to understand for non-chemists and also give those wishing to supply for illegitimate reasons something to 'aim' for in terms of non-controlled chemical space. For both Markush and analogue type approaches, typically computational systems are used to flag likely regulated chemicals.

Criticism

The CSA does not include a definition of "drug abuse". In addition, research shows certain substances on Schedule I, for drugs which have no accepted medical uses and high potential for abuse, actually have accepted medical uses, have low potential for abuse, or both. One of those substances is cannabis, which is either decriminalized or legalized in 33 states of the United States.

8.2 CLASSIFICATION OF CONTROLLED SUBSTANCES

Controlled substances are classified into several schedules based on their potential for abuse, accepted medical use, and safety or dependence liability. The schedules range from Schedule I to Schedule V, with Schedule I substances being the most restricted and Schedule V the least.

8.2.1 Schedule I Substances

Schedule I substances are deemed to have a high potential for abuse and no accepted medical use. Examples include heroin, lysergic acid diethylamide (LSD), and 3,4-methylenedioxymethamphetamine (MDMA or ecstasy). Due to their illicit nature, substances in this category are often the focus of law enforcement investigations, and forensic chemists play a critical role in identifying and quantifying these drugs in seized samples.

8.2.2 Schedule II Substances

Schedule II substances are recognized for having legitimate medical uses but also present a high potential for abuse and dependence. This category includes drugs such as morphine, cocaine, and oxycodone. The regulation of these substances is stringent, and forensic chemistry is essential for ensuring compliance with legal standards regarding their distribution and use.

8.2.3 Schedule III, IV, and V Substances

Schedules III, IV, and V encompass substances with progressively lower potential for abuse. Schedule III substances include anabolic steroids and certain barbiturates; Schedule IV substances feature drugs like diazepam (Valium) and alprazolam (Xanax), while Schedule V substances include medications such as cough preparations containing less than 200 mg of codeine per 100 mL.

Although these drugs are more accessible for medical use, their potential for abuse still necessitates careful regulation and monitoring.

8.3 THE ROLE OF FORENSIC CHEMISTRY

Forensic chemistry is vital in the analysis of controlled substances for various reasons, including crime investigation, regulatory compliance, and public safety.

The discipline provides the necessary techniques to identify, quantify, and characterize these substances accurately, contributing to law enforcement efforts and legal proceedings.

8.3.1 Crime Scene Investigation

In criminal investigations involving controlled substances, forensic chemists analyze samples collected from crime scenes, such as drugs found in possession of suspects, materials seized from drug trafficking operations, and residues found in vehicles or environments. The identification and quantification of these substances can establish links between suspects, locations, and criminal activities, forming a crucial part of the evidence presented in court.

8.3.2 Toxicology Analysis

Forensic toxicology focuses on the effects of drugs and chemicals on the human body. Controlled substances frequently feature in toxicology analyses in cases of overdose, accidental poisoning, or suspected drug-facilitated crimes. By determining the concentration of controlled substances in biological samples (e.g., blood, urine), forensic chemists can provide insights into the cause of death, potential impairment, and the sequence of events leading to a particular incident.

8.3.3 Drug Testing and Regulatory Compliance

Forensic chemistry is also essential in ensuring compliance with regulations governing controlled substances. Laboratories are often tasked with testing pharmaceuticals for purity and compliance with legal standards. Additionally, drug testing programs in workplaces or sports organizations rely on forensic chemistry methods to detect the presence of controlled substances, helping to promote safety and fairness.

8.4 ANALYTICAL METHODS FOR CONTROLLED SUBSTANCES

The identification and quantification of controlled substances involve various analytical techniques, each with its advantages and limitations. Forensic chemists must choose the most appropriate method based on factors such as the nature of the sample, the required sensitivity, and the available resources.

8.4.1 Gas Chromatography (GC)

Gas chromatography is a widely used technique for separating and analyzing volatile substances. In forensic chemistry, GC is often coupled with mass spectrometry (GC-MS) for the identification of controlled substances.

This combination provides high sensitivity and specificity, making it suitable for analyzing complex mixtures, such as those found in seized drugs. GC is particularly effective for analyzing substances like cocaine, amphetamines, and other recreational drugs.

8.4.2 Liquid Chromatography (LC)

Liquid chromatography, including high-performance liquid chromatography (HPLC), is another critical method for analyzing controlled substances. HPLC is particularly useful for separating and quantifying non-volatile and thermally unstable compounds, such as prescription medications and certain synthetic drugs. The versatility of HPLC allows forensic chemists to develop methods tailored to specific substances, enhancing the accuracy of their analyses.

8.4.3 Mass Spectrometry (MS)

Mass spectrometry is a powerful analytical technique that provides detailed information about the molecular weight and structure of compounds. In forensic chemistry, MS can be used in conjunction with various separation methods (such as GC or LC) to confirm the identity of controlled substances and determine their concentrations. This technique is invaluable for identifying new psychoactive substances that may not be included in existing reference databases.

8.4.4 Infrared Spectroscopy (IR)

Infrared spectroscopy is a technique that measures the absorption of infrared light by a sample, providing information about its molecular structure. Forensic chemists use IR spectroscopy to identify controlled substances based on their unique spectral fingerprints. This method is particularly useful for rapid screening of seized substances in the field or laboratory, providing preliminary results before more detailed analyses are performed.

8.5 LEGAL AND ETHICAL CONSIDERATIONS

The regulation of controlled substances is a complex legal framework that varies across jurisdictions. Forensic chemists must adhere to strict protocols and ethical guidelines to ensure the integrity of their analyses and the validity of their results.

8.5.1 Chain of Custody

Maintaining a proper chain of custody is paramount in forensic chemistry. This process involves documenting the handling and transfer of samples from the crime scene to the laboratory and during analysis. A well-maintained chain of custody ensures that the evidence remains uncontaminated and that its integrity is upheld in legal proceedings. Any discrepancies in the chain of custody can undermine the admissibility of evidence in court.

8.5.2 Compliance with Regulations

Forensic laboratories must comply with national and international regulations governing the handling and analysis of controlled substances. These regulations include maintaining appropriate licensing, conducting regular audits, and adhering to strict quality control measures. Forensic chemists must stay informed about changes in legislation related to controlled substances and adapt their practices accordingly.

8.5.3 Ethical Implications

Forensic chemists face ethical considerations in their work, particularly when analyzing controlled substances. They must ensure their analyses are unbiased, accurate, and objective, providing reliable information for legal proceedings. Additionally, forensic chemists must navigate the ethical implications of their findings, especially in cases involving addiction or substance abuse, where their work may impact individuals' lives and communities.

8.6 IMPLICATIONS FOR PUBLIC HEALTH AND SAFETY

The study of controlled substances in forensic chemistry extends beyond criminal justice; it has significant implications for public health and safety. The understanding of drug abuse trends, the development of new psychoactive substances, and the need for effective regulatory measures all highlight the importance of forensic chemistry in safeguarding communities.

8.6.1 Drug Abuse Prevention

Forensic chemistry contributes to efforts in drug abuse prevention by providing data on the prevalence of controlled substances in communities. Understanding drug trends can inform public health campaigns and intervention strategies, helping to address issues related to addiction and overdose.

8.6.2 Response to Emerging Threats

The emergence of new psychoactive substances presents ongoing challenges for forensic chemists and public health officials. Continuous monitoring and analysis of these substances are essential for developing effective detection methods and understanding their effects. Forensic chemists play a critical role in identifying these substances, which can help inform law enforcement and public health responses.

8.6.3 Education and Awareness

Forensic chemists often engage in educational initiatives to raise awareness about the dangers of controlled substances. By collaborating with schools, community organizations, and law enforcement agencies, they can help inform the public about the risks associated with drug use and the importance of seeking help for substance use disorders.

8.7 CONCLUSION

Controlled substances are a significant focus of forensic chemistry, impacting criminal investigations, public health, and regulatory efforts. Through the classification of these substances, the application of advanced analytical techniques, and adherence to legal and ethical standards, forensic chemists contribute to the understanding and management of the challenges posed by controlled substances. As society continues to grapple with issues related to drug abuse and emerging threats, the role of forensic chemistry will remain essential in promoting safety, justice, and public health.

CHAPTER 9
TOXICOLOGY

9.1 UNDERSTANDING THE TOXICOLOGY

Toxicology is a scientific discipline, overlapping with biology, chemistry, pharmacology, and medicine, that involves the study of the adverse effects of chemical substances on living organisms and the practice of diagnosing and treating exposures to toxins and toxicants.

The relationship between dose and its effects on the exposed organism is of high significance in toxicology. Factors that influence chemical toxicity include the dosage, duration of exposure (whether it is acute or chronic), route of exposure, species, age, sex, and environment. Toxicologists are experts on poisons and poisoning. There is a movement for evidence-based toxicology as part of the larger movement towards evidence-based practices. Toxicology is currently contributing to the field of cancer research, since some toxins can be used as drugs for killing tumor cells. One prime example of this is ribosome-inactivating proteins, tested in the treatment of leukemia.

9.1.1 History

Dioscorides, a Greek physician in the court of the Roman emperor Nero, made the first attempt to classify plants according to their toxic and therapeutic effect. A work attributed to the 10th century author Ibn Wahshiyya called the *Book on Poisons* describes various toxic substances and poisonous recipes that can be made using magic. A 14th century Kannada poetic work attributed to the Jain prince Mangarasa, *Khagendra Mani Darpana*, describes several poisonous plants.

The 16th-century Swiss physician Paracelsus is considered "the father" of modern toxicology, based on his rigorous (for the time) approach to understanding the effects of substances on the body. He is credited with the classic toxicology maxim, "*Alle Dinge sind Gift und nichts ist ohne Gift; allein die Dosis macht, dass ein Ding kein Gift ist.*" which translates as, "All things are poisonous and nothing is without poison; only the dose makes a thing not poisonous." This is often condensed to: "The dose makes the poison" or in Latin "Sola dosis facit venenum".

Mathieu Orfila is also considered the modern father of toxicology, having given the subject its first formal treatment in 1813 in his *Traité des poisons*, also called *Toxicologie générale.*

In 1850, Jean Stas became the first person to successfully isolate plant poisons from human tissue. This allowed him to identify the use of nicotine as a poison in the Bocarmé murder case, providing the evidence needed to convict the Belgian Count Hippolyte Visart de Bocarmé of killing his brother-in-law.

9.1.2 Basic Principles

The goal of toxicity assessment is to identify adverse effects of a substance. Adverse effects depend on two main factors: i) routes of exposure (oral, inhalation, or dermal) and ii) dose (duration and concentration of exposure). To explore dose, substances are tested in both acute and chronic models. Generally, different sets of experiments are conducted to determine whether a substance causes cancer and to examine other forms of toxicity.

Factors that influence chemical toxicity:

- **Dosage:**
  - Both large single exposures (acute) and continuous small exposures (chronic) are studied.
- **Route of Exposure:**
  - Ingestion, inhalation or skin absorption
- **Other Factors:**
  - Species
  - Age
  - Sex
  - Health
  - Environment
  - Individual characteristics

The discipline of evidence-based toxicology strives to transparently, consistently, and objectively assess available scientific evidence in order to answer questions in toxicology, the study of the adverse effects of chemical, physical, or biological agents on living organisms and the environment, including the prevention and amelioration of such effects. Evidence-based toxicology has the potential to address concerns in the toxicological community about the limitations of current approaches to assessing the state of the science. These include concerns related to transparency in decision-making, synthesis of different types of evidence, and the assessment of bias and credibility. Evidence-based toxicology has its roots in the larger movement towards evidence-based practices.

9.1.3 Testing Methods

Toxicity experiments may be conducted *in vivo* (using the whole animal) or *in vitro* (testing on isolated cells or tissues), or *in silico* (in a computer simulation).

In Vivo Model Organism

The classic experimental tool of toxicology is testing on non-human animals. Examples of model organisms are *Galleria mellonella,* which can replace small mammals, Zebrafish (*Danio rerio*), which allow for the study of toxicology in a lower order vertebrate *in vivo* and *Caenorhabditis elegans.* As of 2014, such animal testing provides information that is not available by other means about how substances function in a living organism. The use of non-human animals for toxicology testing is opposed by some organisations for reasons of animal welfare, and it has been restricted or banned under some circumstances in certain regions, such as the testing of cosmetics in the European Union.

In Vitro Methods

While testing in animal models remains as a method of estimating human effects, there are both ethical and technical concerns with animal testing. Since the late 1950s, the field of toxicology has sought to reduce or eliminate animal testing under the rubric of "Three Rs" – reduce the number of experiments with animals to the minimum necessary; refine experiments to cause less suffering, and replace *in vivo* experiments with other types, or use more simple forms of life when possible. The historical development of alternative testing methods in toxicology has been published by Balls.

Computer modeling is an example of an alternative in vitro toxicology testing method; using computer models of chemicals and proteins, structure-activity relationships can be determined, and chemical structures that are likely to bind to, and interfere with, proteins with essential functions, can be identified. This work requires expert knowledge in molecular modeling and statistics together with expert judgment in chemistry, biology and toxicology.

In 2007 the American NGO National Academy of Sciences published a report called "Toxicity Testing in the 21st Century: A Vision and a Strategy" which opened with a statement: "Change often involves a pivotal event that builds on previous history and opens the door to a new era.

Pivotal events in science include the discovery of penicillin, the elucidation of the DNA double helix, and the development of computers.... Toxicity testing is approaching such a scientific pivot point. It is poised to take advantage of the revolutions in biology and biotechnology. Advances in toxicogenomics, bioinformatics, systems biology, epigenetics, and computational toxicology could transform toxicity testing from a system based on whole-animal testing to one founded primarily on in vitro methods that evaluate changes in biologic processes using cells, cell lines, or cellular components, preferably of human origin." As of 2014 that vision was still unrealized.

The United States Environmental Protection Agency studied 1,065 chemical and drug substances in their ToxCast program (part of the CompTox Chemicals Dashboard) using *in silica* modelling and a human pluripotent stem cell-based assay to predict *in vivo* developmental intoxicants based on changes in cellular metabolism following chemical exposure.

Major findings from the analysis of this ToxCast_STM dataset published in 2020 include:

- 19% of 1065 chemicals yielded a prediction of developmental toxicity,
- Assay performance reached 79%–82% accuracy with high specificity (> 84%) but modest sensitivity (< 67%) when compared with *in vivo* animal models of human prenatal developmental toxicity,
- Sensitivity improved as more stringent weights of evidence requirements were applied to the animal studies, and
- Statistical analysis of the most potent chemical hits on specific biochemical targets in

ToxCast revealed positive and negative associations with the STM response, providing insights into the mechanistic underpinnings of the targeted endpoint and its biological domain.

In some cases shifts away from animal studies have been mandated by law or regulation; the European Union (EU) prohibited use of animal testing for cosmetics in 2013.

9.1.4 Dose Response Complexities

Most chemicals display a classic dose response curve – at a low dose (below a threshold), no effect is observed. Some show a phenomenon known as sufficient challenge – a small exposure produces animals that "grow more rapidly, have better general appearance and coat quality, have fewer tumors, and live longer than the control animals". A few chemicals have no well-defined safe level of exposure. These are treated with special care. Some chemicals are subject to bioaccumulation as they are stored in rather than being excreted from the body; these also receive special consideration.

Several measures are commonly used to describe toxic dosages according to the degree of effect on an organism or a population, and some are specifically defined by various laws or organizational usage. These include:

- LD50 or LD_{50} = Median lethal dose, a dose that will kill 50% of an exposed population
- NOEL = No-Observed-Effect-Level, the highest dose known to show no effect

- NOAEL = No-Observed-Adverse-Effect-Level, the highest dose known to show no adverse effects
- PEL = Permissible Exposure Limit, the highest concentration permitted under US OSHA regulations
- STEL = Short-Term Exposure Limit, the highest concentration permitted for short periods of time, in general 15–30 minutes
- TWA = Time-Weighted Average, the average amount of an agent's concentration over a specified period of time, usually 8 hours
- TTC = The Threshold of Toxicological Concern concept has been applied to low-level contaminants, such as the constituents of tobacco smoke

9.1.5 Types

- Medical toxicology is the discipline that requires physician status (MD or DO degree plus specialty education and experience).
- Clinical toxicology is the discipline that can be practiced not only by physicians but also other health professionals with a master's degree in clinical toxicology: physician extenders (physician assistants, nurse practitioners), nurses, pharmacists, and allied health professionals.
- Forensic toxicology is the discipline that makes use of toxicology and other disciplines such as analytical chemistry, pharmacology and clinical chemistry to aid medical or legal investigation of death, poisoning, and drug use. The primary concern for forensic toxicology is not the legal outcome of the toxicological investigation or the technology utilized, but rather the obtainment and interpretation of results.
- Computational toxicology is a discipline that develops mathematical and computer-based models to better understand and predict adverse health effects caused by chemicals, such as environmental pollutants and pharmaceuticals. Within the *Toxicology in the 21st Century* project, the best predictive models were identified to be Deep Neural Networks, Random Forest, and Support Vector Machines, which can reach the performance of in vitro experiments.
- Occupational toxicology is the application of toxicology to chemical hazards in the workplace.

9.1.6 Toxicology as a Profession

A toxicologist is a scientist or medical personnel who specializes in the study of symptoms, mechanisms, treatments and detection of venoms and toxins; especially the poisoning of people.

Requirements

To work as a toxicologist one should obtain a degree in toxicology or a related degree like biology, chemistry, pharmacology or biochemistry. Bachelor's degree programs in toxicology cover the chemical makeup of toxins and their effects on biochemistry, physiology and ecology. After introductory life science courses are complete, students typically enroll in labs and apply toxicology principles to research and other studies. Advanced students delve into specific sectors, like the pharmaceutical industry or law enforcement, which apply methods of toxicology in their work. The Society of Toxicology (SOT) recommends that undergraduates in postsecondary schools that do not offer a bachelor's degree in toxicology consider attaining a degree in biology or chemistry. Additionally, the SOT advises aspiring toxicologists to take statistics and mathematics courses, as well as gain laboratory experience through lab courses, student research projects and internships. To become Medical Toxicologists, physicians in the United States complete residency training such as in Emergency Medicine, Pediatrics or Internal Medicine, followed by fellowship in Medical Toxicology and eventual certification by the American College of Medical Toxicology (ACMT).

Duties

Toxicologists perform many different duties including research in the academic, nonprofit and industrial fields, product safety evaluation, consulting, public service and legal regulation. In order to research and assess the effects of chemicals, toxicologists perform carefully designed studies and experiments. These experiments help identify the specific amount of a chemical that may cause harm and potential risks of being near or using products that contain certain chemicals. Research projects may range from assessing the effects of toxic pollutants on the environment to evaluating how the human immune system responds to chemical compounds within pharmaceutical drugs. While the basic duties of toxicologists are to determine the effects of chemicals on organisms and their surroundings, specific job duties may vary based on industry and employment. For example, forensic toxicologists may look for toxic substances in a crime scene, whereas aquatic toxicologists may analyze the toxicity level of water bodies.

Compensation

The salary for jobs in toxicology is dependent on several factors, including level of schooling, specialization, experience. The U.S. Bureau of Labor Statistics (BLS) notes that jobs for biological scientists, which generally include toxicologists, were expected to increase by 21% between 2008 and 2018. The BLS notes that this increase could be due to research and development growth in biotechnology, as well as budget increases for basic and medical research in biological science.

9.2 THE IMPORTANCE OF TOXICOLOGY IN FORENSIC CHEMISTRY

Forensic toxicology serves multiple purposes in the criminal justice system. It is an essential discipline for understanding how substances affect human health and behavior, particularly in cases involving death, injury, or impairment due to chemical exposure.

9.2.1 Investigation of Deaths

One of the primary roles of forensic toxicology is to assist in determining the cause of death in suspicious cases. When a body is discovered under unusual circumstances, toxicologists analyze biological samples such as blood, urine, hair, and tissues to identify the presence of toxic substances, drugs, or alcohol. This analysis can reveal whether the individual died from poisoning, an overdose, or other toxicological causes. By identifying the substances involved, toxicologists can provide critical evidence to law enforcement and medical examiners, aiding in homicide investigations and autopsies.

9.2.2 Drug Impairment Cases

Toxicology also plays a significant role in cases where drug impairment is suspected, such as driving under the influence (DUI) offenses. Forensic toxicologists analyze blood and urine samples from individuals involved in accidents or traffic violations to determine if they were under the influence of drugs or alcohol at the time of the incident. This information is crucial for legal proceedings, as it helps establish culpability and can influence sentencing outcomes.

9.2.3 Poisoning and Chemical Exposure

In cases of suspected poisoning, forensic toxicology is essential for identifying the specific agents involved. Whether the poisoning is intentional or accidental, understanding the chemical nature of the substance and its effects on the body can provide insights into the circumstances surrounding the exposure. Forensic toxicologists often work closely with law enforcement, medical personnel, and public health officials to investigate poisoning incidents, ensuring appropriate measures are taken to protect public health and safety.

9.3 ANALYTICAL METHODS IN FORENSIC TOXICOLOGY

The accurate detection and quantification of toxic substances require a variety of analytical techniques. Forensic toxicologists employ sophisticated methods to analyze samples, each with its strengths and limitations.

9.3.1 Immunoassays

Immunoassays are commonly used as preliminary screening tests for the detection of drugs and their metabolites. These tests utilize antibodies that bind specifically to target substances, allowing for the rapid identification of a wide range of drugs in biological samples. Enzyme-linked immunosorbent assays (ELISA) are particularly popular due to their sensitivity and specificity. However, while immunoassays are effective for screening, they may yield false positives or negatives, necessitating confirmatory testing.

9.3.2 Gas Chromatography-Mass Spectrometry (GC-MS)

Gas chromatography-mass spectrometry (GC-MS) is one of the gold standards in forensic toxicology. This method combines the separation capabilities of gas chromatography with the identification capabilities of mass spectrometry. GC-MS is highly effective for analyzing volatile and semi-volatile substances, making it suitable for detecting drugs, alcohol, and their metabolites in complex biological matrices. The specificity and sensitivity of GC-MS make it an essential tool for confirmatory testing, ensuring accurate results in forensic investigations.

9.3.3 Liquid Chromatography-Mass Spectrometry (LC-MS)

Liquid chromatography-mass spectrometry (LC-MS) is another powerful analytical technique widely used in forensic toxicology. Unlike GC-MS, LC-MS is capable of analyzing non-volatile and thermally labile compounds, allowing toxicologists to detect a broader range of substances, including certain prescription medications and synthetic drugs. The versatility of LC-MS has made it an invaluable tool in forensic laboratories, especially as new psychoactive substances continue to emerge.

9.3.4 Solid-Phase Microextraction (SPME)

Solid-phase microextraction (SPME) is an innovative technique that simplifies the sample preparation process while providing efficient extraction of analytes. SPME uses a fiber coated with a sorbent material to absorb volatile and semi-volatile compounds from samples, which can then be analyzed using techniques such as GC-MS or LC-MS. This method enhances the sensitivity of analyses and reduces the time and effort required for sample preparation, making it particularly useful in forensic toxicology.

9.4 CHALLENGES IN FORENSIC TOXICOLOGY

While forensic toxicology plays a vital role in the criminal justice system, practitioners face several challenges that can complicate their work.

9.4.1 Emerging Drugs and Substances

The continuous emergence of new psychoactive substances poses a significant challenge for forensic toxicologists. Many of these synthetic drugs are designed to evade detection by standard analytical methods, making it difficult for toxicologists to identify and quantify them. As new substances enter the market, forensic toxicologists must stay updated on the latest trends and develop new analytical methods to detect these compounds effectively.

9.4.2 Matrix Complexity

Biological samples, such as blood, urine, and hair, can be complex matrices that contain a wide variety of substances, including metabolites and impurities. This complexity can hinder the accurate detection and quantification of target analytes. Forensic toxicologists must carefully select and optimize their analytical methods to account for potential interferences and ensure reliable results.

9.4.3 Legal and Ethical Considerations

Forensic toxicologists must navigate various legal and ethical considerations in their work. They are often required to adhere to strict protocols and regulations governing the handling and analysis of biological samples. Additionally, toxicologists must ensure that their findings are communicated effectively to law enforcement, legal professionals, and the courts, maintaining transparency and accuracy throughout the process. Any discrepancies or misinterpretations of toxicological evidence can have significant implications for legal outcomes.

9.5 THE ROLE OF FORENSIC TOXICOLOGISTS IN THE LEGAL SYSTEM

Forensic toxicologists play a crucial role in bridging the gap between science and the legal system. Their expertise is essential in providing clear and unbiased analyses that can withstand scrutiny in court.

9.5.1 Expert Testimony

In legal proceedings, forensic toxicologists often serve as expert witnesses, providing testimony about their analyses and findings. They must communicate complex scientific concepts in a manner that is understandable to judges, juries, and attorneys. This requires not only a thorough understanding of toxicological principles but also the ability to articulate the significance of their findings within the context of the case.

9.5.2 Collaboration with Law Enforcement and Legal Professionals

Forensic toxicologists frequently collaborate with law enforcement agencies and legal professionals throughout the investigation process. They provide guidance on the collection and handling of biological samples, ensuring that evidence is preserved and appropriately analyzed. By working closely with investigators, toxicologists can help formulate strategies for obtaining and interpreting evidence related to drug use, poisoning, and other toxicological issues.

9.5.3 Public Health Implications

The work of forensic toxicologists extends beyond individual cases; it also has broader implications for public health and safety. By identifying trends in drug use and poisoning incidents, toxicologists can contribute valuable information to public health initiatives and policy development. Their findings can inform educational campaigns, substance abuse prevention programs, and regulatory efforts aimed at addressing the challenges posed by controlled substances and toxic exposures.

9.6 CONCLUSION

Toxicology is a fundamental aspect of forensic chemistry that plays a critical role in the investigation of crimes and the legal system. Through various analytical methods, forensic toxicologists can identify and quantify toxic substances in biological samples, providing essential information for criminal investigations, public health efforts, and legal proceedings. Despite the challenges posed by emerging drugs, complex matrices, and legal considerations, forensic toxicologists continue to advance the field through innovation and collaboration. Their work not only aids in the pursuit of justice but also contributes to the ongoing efforts to promote public health and safety in our communities. As the landscape of toxicology evolves, the expertise and contributions of forensic toxicologists will remain indispensable in addressing the complexities of chemical exposure and substance abuse.

CHAPTER 10
TRACE EVIDENCE

10.1 UNDERSTANDING THE TRACE EVIDENCE

Trace evidence refers to small, often microscopic materials transferred during a crime. These materials can include hair, fibers, paint, glass, soil, gunshot residue, and other tiny particles that are unintentionally left behind by individuals or objects involved in criminal activity. Trace evidence plays a crucial role in forensic investigations, as it can help establish connections between suspects, victims, and crime scenes.

Although small in size, trace evidence can be incredibly powerful in corroborating other forms of evidence or even serving as a primary link in solving a case. Trace evidence, though often microscopic, can be a powerful tool in forensic investigations. By linking individuals to crime scenes and supporting other forms of evidence, it can provide valuable insights into what occurred during a crime. As forensic science continues to advance, the techniques for collecting, analyzing, and interpreting trace evidence will only become more refined, enhancing its role in the pursuit of justice.

10.1.1 Collection and Preservation of Trace Evidence

The proper collection and preservation of trace evidence are critical to ensuring its integrity for forensic analysis. Crime scene investigators are trained to handle trace evidence carefully to avoid contamination or loss. Trace evidence is often small, making it vulnerable to being overlooked or destroyed.

Collection Techniques

There are various methods for collecting trace evidence, depending on its type. Common techniques include:

- **Taping**: For collecting hair, fibers, and other particles on surfaces or clothing, adhesive tapes are used to lift evidence gently.
- **Vacuuming**: For large areas, such as carpets or car interiors, vacuuming with special filters can collect a variety of trace materials.
- **Swabbing**: Gunshot residue and other small particles can be collected using swabs, which are gently rubbed over the suspect's hands or clothing.

- **Scraping:** Tools like scalpels are used to scrape off paint or other materials from surfaces where trace evidence might be found.

Preservation Methods

Once collected, trace evidence must be carefully stored to prevent contamination or degradation. Evidence is typically sealed in tamper-proof containers, labeled, and stored in a secure environment. Each item is documented to maintain a clear chain of custody. In some cases, evidence may need to be stored in climate-controlled environments, especially organic materials like hair or plant matter, to prevent decay.

10.2 THE IMPORTANCE OF TRACE EVIDENCE IN FORENSIC CHEMISTRY

Trace evidence is invaluable in forensic investigations for several reasons. It often serves as a link between the perpetrator and the crime scene, making it possible to reconstruct events and establish connections among individuals, locations, and objects involved in the crime.

10.2.1 Connecting Suspects to Crime Scenes

One of the most significant roles of trace evidence is its ability to establish links between suspects and crime scenes. For example, fibers from a suspect's clothing can be found on a victim, indicating direct contact. Similarly, paint chips from a vehicle can be transferred to the scene of a hit-and-run accident, providing vital clues that help investigators identify the vehicle involved. This type of evidence can be crucial in corroborating witness statements and building a case against a suspect.

10.2.2 Identifying Victims

Trace evidence can also play a critical role in identifying victims, particularly in cases involving decomposed or burned bodies. Forensic chemists can analyze trace materials found on or around a victim to provide insights into their identity. For instance, soil samples from a crime scene can be compared to soil from a suspect's property, helping to establish a connection. In cases where traditional identification methods are not viable, trace evidence may offer the only means of determining a victim's identity.

10.2.3 Supporting or Refuting Claims

In forensic investigations, trace evidence can either support or refute the claims made by suspects, witnesses, or victims. For instance, the presence of gunshot residue on a suspect's hands may support allegations of involvement in a shooting, while the absence of such evidence can be crucial in defending against those charges. Similarly, trace evidence can provide context about the crime, helping to clarify the sequence of events.

10.3 TYPES OF TRACE EVIDENCE

Several types of trace evidence are commonly encountered in forensic investigations, each with unique characteristics and significance.

10.3.1 Fibers

Fibers are one of the most common forms of trace evidence. They can originate from clothing, carpets, upholstery, or other textile materials. Forensic chemists analyze fibers to determine their composition, color, and other characteristics, allowing them to establish potential sources.

The transfer of fibers can occur through direct contact, making them a valuable link between suspects and victims.

Forensic analysts often compare fibers found at a crime scene with known samples to determine similarities and make inferences about the relationships between individuals and locations.

10.3.2 Paint

Paint can also serve as trace evidence in criminal investigations. It can be transferred during hit-and-run incidents, burglaries, or vandalism. Forensic chemists analyze paint samples to identify their composition, color, and type, allowing them to match them to specific vehicles or structures. The examination of paint layers and chemical constituents can provide insights into how and when the paint was applied, helping to establish timelines in investigations.

10.3.3 Glass

Glass fragments found at crime scenes can provide essential information regarding the events that occurred. Forensic chemists analyze glass to determine its type, color, and thickness, allowing them to match fragments to specific sources, such as windows or bottles. The analysis of glass fracture patterns can also provide insights into the circumstances surrounding a crime, such as whether a window was broken from the inside or outside.

10.3.4 Soil and Dust

Soil and dust can be significant trace evidence, especially in outdoor crime scenes. Forensic chemists analyze soil samples to determine their mineral content, pH, and organic material, helping to establish geographic connections between crime scenes and suspects. Dust samples can also provide valuable information about a suspect's environment and activities, linking them to specific locations.

10.3.5 Biological Evidence

Biological materials, such as hair, skin cells, or bodily fluids, can also be considered trace evidence. Forensic chemists analyze these samples using various techniques to determine their origin, which can provide critical insights into the relationships between individuals and crime scenes. Hair samples can be examined for mitochondrial DNA analysis, which may link a suspect to a victim or location.

10.4 ANALYTICAL TECHNIQUES FOR TRACE EVIDENCE

The examination of trace evidence involves various analytical techniques, each with its strengths and limitations. Forensic chemists employ these methods to accurately analyze and interpret trace materials.

10.4.1 Microscopy

Microscopic analysis is a fundamental technique used in the examination of trace evidence. Forensic scientists use various types of microscopes, including comparison microscopes, to analyze fibers, hair, and other small materials. Microscopy allows forensic chemists to observe characteristics such as color, texture, and shape, enabling them to compare unknown samples with known references.

10.4.2 Spectroscopy

Spectroscopic techniques, such as infrared (IR) spectroscopy and Raman spectroscopy, are often used to analyze the chemical composition of trace evidence. These techniques provide valuable information about the molecular structure and chemical properties of materials, allowing forensic chemists to identify unknown substances. For instance, IR spectroscopy can be used to analyze fibers and paint samples, while Raman spectroscopy can be applied to analyze glass and other materials.

10.4.3 Chromatography

Chromatography is another essential analytical technique employed in forensic chemistry. Gas chromatography (GC) and high-performance liquid chromatography (HPLC) are commonly used to separate and analyze complex mixtures of substances. For instance, GC can be used to analyze volatile compounds in paint samples, while HPLC is effective for analyzing dyes in fibers. These techniques help forensic chemists identify specific components within trace evidence, providing crucial information for investigations.

10.4.4 Mass Spectrometry

Mass spectrometry (MS) is often coupled with chromatography to provide detailed information about the composition of trace evidence. This powerful analytical technique allows forensic chemists to determine the molecular weight and structure of compounds, enabling them to identify unknown substances with high accuracy. By combining MS with GC or HPLC, forensic scientists can obtain comprehensive profiles of trace evidence, aiding in investigations.

10.5 CHALLENGES IN ANALYZING TRACE EVIDENCE

While trace evidence is a powerful tool in forensic investigations, forensic chemists face several challenges that can complicate the analysis process.

10.5.1 Transfer and Contamination

One of the primary challenges in trace evidence analysis is the potential for transfer and contamination.

Trace evidence can easily be lost, altered, or contaminated during the collection and handling process. Proper procedures must be followed to ensure that samples are collected, preserved, and analyzed without introducing extraneous materials. Contamination can compromise the integrity of the evidence and lead to inaccurate conclusions.

10.5.2 Interpretation of Results

The interpretation of trace evidence can be subjective, as it often relies on the experience and expertise of the forensic chemist. Establishing the significance of a match between trace evidence and known samples can be complex, particularly in cases involving common materials or environmental factors.

Forensic chemists must be cautious in their conclusions and communicate their findings clearly to avoid misinterpretation in legal contexts.

10.5.3 Emerging Technologies

As new technologies and materials emerge, forensic chemists must continuously adapt their methodologies to keep pace with these changes. The development of synthetic fibers, new paint formulations, and advanced manufacturing processes can complicate the identification and analysis of trace evidence.

Forensic chemists must remain informed about the latest trends in materials science to ensure that their analyses are relevant and effective.

10.6 Conclusion

Trace evidence is a critical component of forensic chemistry, providing essential links between suspects, victims, and crime scenes. Through the analysis of various types of trace evidence, forensic chemists can help reconstruct events, establish connections, and support legal proceedings.

The analytical techniques employed in trace evidence analysis, including microscopy, spectroscopy, chromatography, and mass spectrometry, enable forensic scientists to obtain detailed insights into the nature of the materials involved. Despite the challenges faced in the field, the significance of trace evidence in forensic investigations cannot be overstated.

As forensic science continues to evolve, the role of trace evidence in unraveling the complexities of criminal cases will remain indispensable in the pursuit of justice.

CHAPTER 11

QUESTIONED DOCUMENTS AND IMPRESSION EVIDENCE

11.1 QUESTIONED DOCUMENTS AND IMPRESSION EVIDENCE

Forensic chemistry encompasses a broad spectrum of disciplines, one of which is the analysis of questioned documents and impression evidence. These areas are pivotal in forensic investigations, aiding in the identification of forged documents, the examination of handwriting, and the analysis of physical impressions left by objects.

This essay explores the significance of questioned documents and impression evidence in forensic chemistry, the methods employed in their analysis, and their implications in criminal investigations.

11.2 UNDERSTANDING QUESTIONED DOCUMENTS

Questioned documents refer to any handwritten, typed, printed, or otherwise prepared material whose authenticity or origin is disputed. Such documents can include wills, contracts, identification cards, checks, and even handwritten notes. The analysis of questioned documents plays a crucial role in forensic investigations by providing insights into the legitimacy of documents and the identities of individuals involved.

11.2.1 Types of Questioned Documents

There are various types of questioned documents, each presenting unique challenges for forensic analysis. Handwritten documents are perhaps the most common form of questioned documents. Forensic experts analyze the characteristics of handwriting, such as letter formation, slant, pressure, and spacing, to determine whether the writing matches a known sample from a suspect or individual of interest.

Typed documents can also be questioned, especially if there is suspicion of forgery or tampering. Forensic analysis may involve examining typewriter characteristics, font discrepancies, or the presence of specific artifacts indicative of digital manipulation in electronic documents. Additionally, documents containing alterations, erasures, or obliterations present unique challenges, as forensic chemists must use specialized techniques to reveal the original content.

11.2.2 Significance in Forensic Investigations

The analysis of questioned documents is essential for several reasons. First, it helps establish the authenticity of a document, which can be critical in legal disputes, such as contested wills or fraudulent contracts. Second, it can link individuals to criminal activities by confirming the authorship of threatening letters, ransom notes, or other communications. Lastly, the examination of questioned documents can aid in understanding the context of a crime, providing insights into the motives and actions of individuals involved.

11.3 IMPRESSION EVIDENCE

Impression evidence refers to marks left by objects when they come into contact with a surface. This type of evidence is crucial in forensic investigations, as it can link suspects to crimes or provide insights into how a crime was committed. Impression evidence can include shoeprints, tire treads, tool marks, and bite marks, all of which offer valuable information for forensic chemists.

11.3.1 Types of Impression Evidence

- **Shoeprints:** Shoeprints are among the most common types of impression evidence. They can reveal the brand, model, and size of the shoe, helping to identify potential suspects. Forensic experts analyze the tread patterns and any unique features of the shoeprint, such as wear patterns or defects, to establish a match with a suspect's footwear.
- **Tire Treads:** Similar to shoeprints, tire tread impressions can provide critical information in vehicular-related crimes. Forensic analysis of tire treads can reveal the make and model of a vehicle, as well as its size and type. By comparing tread patterns from crime scenes with known tire samples, investigators can link a vehicle to a crime.
- **Tool Marks:** Tool marks are impressions left by tools when they come into contact with a surface, such as a door frame or window. Forensic analysis of tool marks can help identify the specific tool used in a crime, providing valuable information about the perpetrator's methods. Forensic chemists examine the unique characteristics of tool marks, such as striations and impressions, to establish a match.
- **Bite Marks:** Bite marks can provide forensic evidence in cases involving assault or sexual offenses. Forensic experts analyze the size, shape, and arrangement of bite marks to compare them with dental impressions of potential suspects. While the analysis of bite marks is complex and often controversial, it can provide valuable information in certain cases.

11.3.2 The Role of Forensic Chemistry in Impression Evidence

Forensic chemistry plays a critical role in the analysis of impression evidence. Forensic chemists employ various techniques to enhance and preserve impressions for analysis. For example, dust prints can be made visible using powders or electrostatic lifting techniques, while latent impressions can be enhanced using specialized chemicals or lighting techniques. Once the impressions are visible, forensic experts can take photographs and create casts for further analysis.

The comparison of impression evidence requires a meticulous approach. Forensic chemists analyze unique features in the impressions and compare them to known samples from suspects or objects of interest. The process involves assessing factors such as size, shape, depth, and unique characteristics to establish a match. In some cases, advanced imaging techniques or 3D scanning may be employed to facilitate comparisons.

11.4 METHODS OF ANALYSIS

The analysis of questioned documents and impression evidence involves a combination of visual examination, chemical analysis, and comparison techniques.

11.4.1 Visual Examination

The initial step in analyzing questioned documents involves a careful visual examination. Forensic experts look for signs of alterations, such as erasures, overwriting, or ink differences.

The analysis may also include the assessment of the paper type, printing methods, and any watermarks or security features that can provide insights into the document's authenticity.

11.4.2 Chemical Analysis

In cases where ink or writing materials are questioned, forensic chemists may conduct chemical analyses. Techniques such as thin-layer chromatography (TLC) can separate ink components, allowing experts to compare inks from questioned documents with known samples. This analysis can help establish whether the same ink was used in the creation of the document in question.

11.4.3 Comparison Techniques

For both questioned documents and impression evidence, comparison techniques are paramount. In the case of handwriting analysis, forensic experts use known samples to compare letter formations, slant, and pressure. When analyzing impressions, forensic chemists use known footwear or tire samples to compare tread patterns and other features. Advanced imaging techniques, such as digital comparison software, can enhance the accuracy of these comparisons.

11.5 CHALLENGES IN ANALYZING QUESTIONED DOCUMENTS AND IMPRESSION EVIDENCE

The analysis of questioned documents and impression evidence presents several challenges that forensic chemists must navigate.

11.5.1 Subjectivity in Analysis

One of the primary challenges in the analysis of questioned documents and impressions is the inherent subjectivity involved. Handwriting analysis, for example, relies on the expertise and experience of the forensic examiner, leading to potential biases or differing interpretations. To address this issue, it is essential for forensic chemists to adhere to standardized protocols and collaborate with other experts to ensure the reliability of their findings.

11.5.2 Quality of Evidence

The quality of questioned documents and impression evidence can vary significantly. Poorly preserved or damaged documents may present challenges in analysis, making it difficult to draw conclusive results. Similarly, impressions that are faint or contaminated may require advanced techniques for enhancement, which may not always yield definitive results. Ensuring proper collection, preservation, and analysis of evidence is crucial for obtaining reliable outcomes.

11.5.3 Legal Considerations

In forensic investigations, the findings from questioned documents and impression evidence analyses can have significant legal implications. It is essential for forensic chemists to communicate their findings clearly and accurately to law enforcement and legal professionals. Additionally, experts must be prepared to testify in court, explaining the methods used in their analyses and the significance of their findings.

11.6 CONCLUSION

Questioned documents and impression evidence are vital components of forensic chemistry, providing critical insights into criminal investigations. The analysis of these materials aids in establishing authenticity, linking suspects to crimes, and understanding the context of criminal activities. By employing various analytical techniques, forensic chemists can uncover vital information that supports investigations and contributes to the pursuit of justice. Despite the challenges associated with analyzing questioned documents and impressions, the significance of this work in forensic investigations cannot be overstated. As forensic science continues to evolve, the importance of questioned documents and impression evidence will remain a cornerstone of criminal investigations, helping to unravel complex cases and bring offenders to justice.

CHAPTER 12
LATENT PRINT DEVELOPMENT

Latent print development is a critical component of forensic chemistry that plays an essential role in criminal investigations. Latent prints, often referred to as fingerprints, are impressions left behind on surfaces by the natural oils and sweat from the skin.

These prints are not visible to the naked eye but can be developed through various chemical and physical methods, allowing forensic experts to identify individuals and link them to crime scenes.

12.1 FINGERPRINT

A fingerprint is an impression left by the friction ridges of a human finger. The recovery of partial fingerprints from a crime scene is an important method of forensic science. Moisture and grease on a finger result in fingerprints on surfaces such as glass or metal.

Deliberate impressions of entire fingerprints can be obtained by ink or other substances transferred from the peaks of friction ridges on the skin to a smooth surface such as paper. Fingerprint records normally contain impressions from the pad on the last joint of fingers and thumbs, though fingerprint cards also typically record portions of lower joint areas of the fingers.

Human fingerprints are detailed, unique, difficult to alter, and durable over the life of an individual, making them suitable as long-term markers of human identity.

They may be employed by police or other authorities to identify individuals who wish to conceal their identity, or to identify people who are incapacitated or deceased and thus unable to identify themselves, as in the aftermath of a natural disaster.

Their use as evidence has been challenged by academics, judges and the media. There are no uniform standards for point-counting methods, and academics have argued that the error rate in matching fingerprints has not been adequately studied and that fingerprint evidence has no secure statistical foundation.

Research has been conducted into whether experts can objectively focus on feature information in fingerprints without being misled by extraneous information, such as context.

Fig 1: A Fingerprint

12.1.1 History

Antiquity and the Medieval Period

Fingerprints have been found on ancient clay tablets, seals, and pottery. They have also been found on the walls of Egyptian tombs and on Minoan, Greek, and Chinese pottery. In ancient China officials authenticated government documents with their fingerprints. In about 200 BC, fingerprints were used to sign written contracts in Babylon. Fingerprints from 3D-scans of cuneiform tablets are extracted using the GigaMesh Software Framework.

With the advent of silk and paper in China, parties to a legal contract impressed their handprints on the document. Sometime before 851 CE, an Arab merchant in China, Abu Zayd Hasan, witnessed Chinese merchants using fingerprints to authenticate loans.

References from the age of the Babylonian king Hammurabi (reigned 1792–1750 BCE) indicate that law officials would take the fingerprints of people who had been arrested. During China's Qin dynasty, records have shown that officials took hand prints and foot prints as well as fingerprints as evidence from a crime scene. In 650, the Chinese historian Kia Kung-Yen remarked that fingerprints could be used as a means of authentication. In his *Jami al-Tawarikh* (Universal History), the Iranian physician Rashid-al-Din Hamadani (1247–1318) refers to the Chinese practice of identifying people via their fingerprints, commenting: "Experience shows that no two individuals have fingers exactly alike." Whether these examples indicate that ancient peoples realized that fingerprints could uniquely identify individuals has been debated, with some arguing these examples are no more meaningful than an illiterate's mark on a document or an accidental remnant akin to a potter's mark on their clay.

Europe in the 17th and 18th Centuries

From the late 16th century onwards, European academics attempted to include fingerprints in scientific studies. But plausible conclusions could be established only from the mid-17th century onwards. In 1686, the professor of anatomy at the University of Bologna Marcello Malpighi identified ridges, spirals and loops in fingerprints left on surfaces. In 1788, a German anatomist Johann Christoph Andreas Mayer was the first European to conclude that fingerprints were unique to each individual.

19th Century

Fig 2: Nine Fingerprint Patterns Identified by Jan Evangelista Purkyne

In 1823, Jan Evangelista Purkyně identified nine fingerprint patterns. The nine patterns include the tented arch, the loop, and the whorl, which in modern-day forensics are considered ridge details. In 1840, following the murder of Lord William Russell, a provincial doctor, Robert Blake Overton, wrote to Scotland Yard suggesting checking for fingerprints. In 1853, the German anatomist Georg von Meissner (1829–1905) studied friction ridges, and in 1858, Sir William James

Herschel initiated fingerprinting in India. In 1877, he first instituted the use of fingerprints on contracts and deeds to prevent the repudiation of signatures in Hooghly near Kolkata and he registered government pensioners' fingerprints to prevent the collection of money by relatives after a pensioner's death.

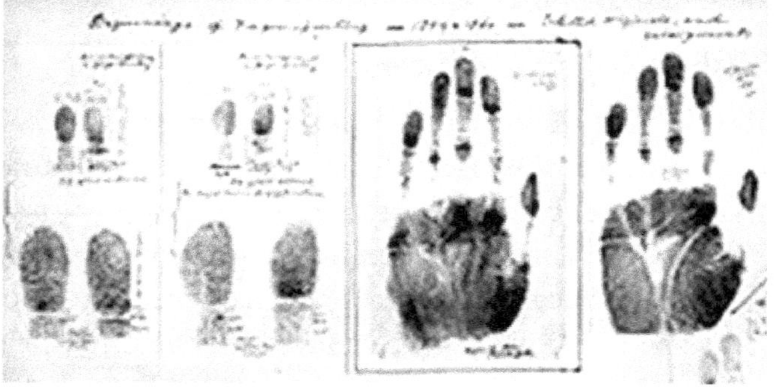

Fig 3: Fingerprints taken by William Herschel 1859/60

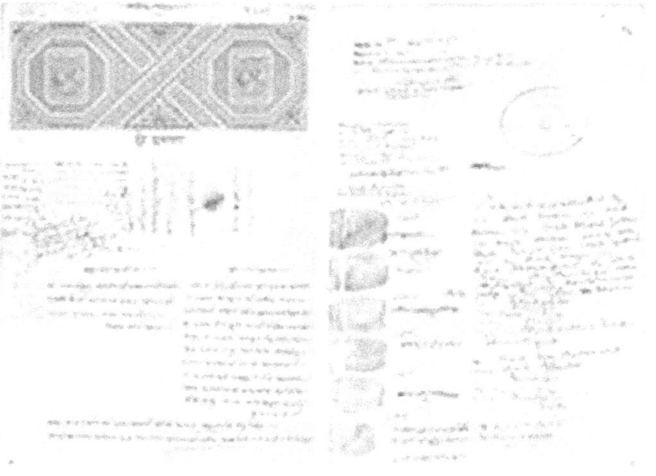

Fig 4: Fingerprints Used Instead of Signatures on an Indian Legal Document of 1952

In 1880, Henry Faulds, a Scottish surgeon in a Tokyo hospital, published his first paper on the usefulness of fingerprints for identification and proposed a method to record them with printing ink. Henry Faulds also suggested, based on his studies, that fingerprints are unique to a human. Returning to Great Britain in 1886, he offered the concept to the Metropolitan Police in London but it was dismissed at that time. Up until the early 1890s, police forces in the United States and on the European continent could not reliably identify criminals to track their criminal record. Francis Galton published a detailed statistical model of fingerprint analysis and identification in his 1892 book *Finger Prints*. He had calculated that the chance of a "false positive" (two different individuals having the same fingerprints) was about 1 in 64 billion. In 1892, Juan Vucetich, an Argentine chief police officer, created the first method of recording the fingerprints of individuals on file. In that same year, Francisca Rojas was found in a house with neck injuries, while her two sons were found dead with their throats cut. Rojas accused a neighbour, but despite brutal interrogation, this neighbour would not confess to the crimes. Inspector Álvarez, a colleague of Vucetich, went to the scene and found a bloody thumb mark on a door. When it was compared with Rojas' prints, it was found to be identical with her right thumb. She then confessed to the murder of her sons. This was the first known murder case to be solved using fingerprint analysis.

In Kolkata, a fingerprint Bureau was established in 1897, after the Council of the Governor General approved a committee report that fingerprints should be used for the classification of criminal records. The bureau employees Azizul Haque and Hem Chandra Bose have been credited with the primary development of a fingerprint classification system eventually named after their supervisor, Sir Edward Richard Henry.

20th Century

The French scientist Paul-Jean Coulier developed a method to transfer latent fingerprints on surfaces to paper using iodine fuming. It allowed the London Scotland Yard to start fingerprinting individuals and identify criminals using fingerprints in 1901. Soon after, American police departments adopted the same method and fingerprint identification became a standard practice in the United States.

The Scheffer case of 1902 is the first case of the identification, arrest, and conviction of a murderer based upon fingerprint evidence. Alphonse Bertillon identified the thief and murderer Scheffer, who had previously been arrested and his fingerprints filed some months before, from the fingerprints found on a fractured glass showcase, after a theft in a dentist's apartment where the dentist's employee was found dead. It was able to be proved in court that the fingerprints had been made after the showcase was broken.

The identification of individuals through fingerprints for law enforcement has been considered essential in the United States since the beginning of the 20th century. Body identification using fingerprints has also been valuable in the aftermath of natural disasters and anthropogenic hazards. In the United States, the FBI manages a fingerprint identification system and database called the Integrated Automated Fingerprint Identification System (IAFIS), which currently holds the fingerprints and criminal records of over 51 million criminal record subjects and over 1.5 million civil (non-criminal) fingerprint records. OBIM, formerly U.S. VISIT, holds the largest repository of biometric identifiers in the U.S. government at over 260 million individual identities. When it was deployed in 2004, this repository, known as the

Automated Biometric Identification System (IDENT), stored biometric data in the form of two-finger records. Between 2005 and 2009, the DHS transitioned to a ten-print record standard in order to establish interoperability with IAFIS.

In 1910, Edmond Locard established the first forensic lab in France. Criminals may wear gloves to avoid leaving fingerprints. However, the gloves themselves can leave prints that are as unique as human fingerprints. After collecting glove prints, law enforcement can match them to gloves that they have collected as evidence or to prints collected at other crime scenes. In many jurisdictions the act of wearing gloves itself while committing a crime can be prosecuted as an inchoate offense.

12.1.2 Biology

Fingerprints are impressions left on surfaces by the friction ridges on the finger of a human. The matching of two fingerprints is among the most widely used and most reliable biometric techniques. Fingerprint matching considers only the obvious features of a fingerprint.

The composition of fingerprints consists of water (95%-99%), as well as organic and inorganic constituents. The organic component is made up of amino acids, proteins, glucose, lactase, urea, pyruvate, fatty acids and sterols. Inorganic ions such as chloride, sodium, potassium and iron are also present. Other contaminants such as oils found in cosmetics, drugs and their metabolites and food residues may be found in fingerprint residues.

A friction ridge is a raised portion of the epidermis on the digits (fingers and toes), the palm of the hand or the sole of the foot, consisting of one or more connected ridge units of friction ridge skin.

These are sometimes known as "epidermal ridges" which are caused by the underlying interface between the dermal papillae of the dermis and the interpapillary (rete) pegs of the epidermis.

These unique features are formed at around the 15th week of fetal development and remain until after death, when decomposition begins. During the development of the fetus, around the 13th week of a pregnancy, ledge-like formation is formed at the bottom of the epidermis beside the dermis. The cells along these ledges begin to rapidly proliferate. This rapid proliferation forms primary and secondary ridges. Both the primary and secondary ridges act as a template for the outer layer of the skin to form the friction ridges seen on the surface of the skin.

These epidermal ridges serve to amplify vibrations triggered, for example, when fingertips brush across an uneven surface, better transmitting the signals to sensory nerves involved in fine texture perception. These ridges may also assist in gripping rough surfaces and may improve surface contact in wet conditions.

12.1.3 Genetics

Consensus within the scientific community suggests that the dermatoglyphic patterns on fingertips are hereditary. The fingerprint patterns between monozygotic twins have been shown to be very similar (though not identical), whereas dizygotic twins have considerably less similarity. Significant heritability has been identified for 12 dermatoglyphic characteristics. Current models of dermatoglyphic trait inheritance suggest Mendelian transmission with additional effects from either additive or dominant major genes.

Whereas genes determine the general characteristics of patterns and their type, the presence of environmental factors result in the slight differentiation of each fingerprint. However, the relative influences of genetic and environmental effects on fingerprint patterns are generally unclear. One study has suggested that roughly 5% of the total variability is due to small environmental effects, although this was only performed using total ridge count as a metric. Several models of finger ridge formation mechanisms that lead to the vast diversity of fingerprints have been proposed. One model suggests that a buckling instability in the basal cell layer of the fetal epidermis is responsible for developing epidermal ridges. Additionally, blood vessels and nerves may also serve a role in the formation of ridge configurations. Another model indicates that changes in amniotic fluid surrounding each developing finger within the uterus cause corresponding cells on each fingerprint to grow in different microenvironments. For a given individual, these various factors affect each finger differently, preventing two fingerprints from being identical while still retaining similar patterns.

It is important to note that the determination of fingerprint inheritance is made difficult by the vast diversity of phenotypes. Classification of a specific pattern is often subjective (lack of consensus on the most appropriate characteristic to measure quantitatively) which complicates analysis of dermatoglyphic patterns. Several modes of inheritance have been suggested and observed for various fingerprint patterns. Total fingerprint ridge count, a commonly used metric of fingerprint pattern size, has been suggested to have a polygenic mode of inheritance and is influenced by multiple additive genes. This hypothesis has been challenged by other research, however, which indicates that ridge counts on individual fingers are genetically independent and lack evidence to support the existence of additive genes influencing pattern formation. Another mode of fingerprint pattern inheritance suggests that the arch pattern on the thumb and on other fingers are inherited as an autosomal dominant trait. Further research on the arch pattern has suggested that a major gene or multifactorial inheritance is responsible for arch pattern heritability. A separate model for the development of the whorl pattern indicates that a single gene or group of linked genes contributes to its inheritance.

Furthermore, inheritance of the whorl pattern does not appear to be symmetric in that the pattern is seemingly randomly distributed among the ten fingers of a given individual. In general, comparison of fingerprint patterns between left and right hands suggests an asymmetry in the effects of genes on fingerprint patterns, although this observation requires further analysis.

In addition to proposed models of inheritance, specific genes have been implicated as factors in fingertip pattern formation (their exact mechanism of influencing patterns is still under research). Multivariate linkage analysis of finger ridge counts on individual fingers revealed linkage to chromosome 5q14.1 specifically for the ring, index, and middle fingers. In mice, variants in the gene EVI1 were correlated with dermatoglyphic patterns. EVI1 expression in humans does not directly influence fingerprint patterns but does affect limb and digit formation which in turn may play a role in influencing fingerprint patterns. Genome-wide association studies found single nucleotide polymorphisms within the gene ADAMTS9-AS2 on 3p14.1, which appeared to have an influence on the whorl pattern on all digits. This gene encodes antisense RNA which may inhibit ADAMTS9, which is expressed in the skin. A model of how genetic variants of ADAMTS9-AS2 directly influence whorl development has not yet been proposed.

In February 2023, a study identified WNT, BMP and EDAR as signaling pathways regulating the formation of primary ridges on fingerprints, with the first two having an opposite relationship established by a Turing reaction-diffusion system.

12.1.4 Classification Systems

Before computerization, manual filing systems were used in large fingerprint repositories. A fingerprint classification system groups fingerprints according to their characteristics and therefore helps in the matching of a fingerprint against a large database of fingerprints.

A query fingerprint that needs to be matched can therefore be compared with a subset of fingerprints in an existing database. Early classification systems were based on the general ridge patterns, including the presence or absence of circular patterns, of several or all fingers. This allowed the filing and retrieval of paper records in large collections based on friction ridge patterns alone. The most popular systems used the pattern class of each finger to form a numeric key to assist lookup in a filing system. Fingerprint classification systems included the Roscher System, the Juan Vucetich System and the Henry Classification System. The Roscher System was developed in Germany and implemented in both Germany and Japan. The Vucetich System was developed in Argentina and implemented throughout South America. The Henry Classification System was developed in India and implemented in most English-speaking countries.

In the Henry Classification System, there are three basic fingerprint patterns: loop, whorl, and arch, which constitute 60–65 percent, 30–35 percent, and 5 percent of all fingerprints respectively. There are also more complex classification systems that break down patterns even further, into plain arches or tented arches, and into loops that may be radial or ulnar, depending on the side of the hand toward which the tail points. Ulnar loops start on the pinky-side of the finger, the side closer to the ulna, the lower arm bone. Radial loops start on the thumb-side of the finger, the side closer to the radius. Whorls may also have sub-group classifications including plain whorls, accidental whorls, double loop whorls, peacock's eye, composite, and central pocket loop whorls.

The "primary classification number" in the Henry Classification System is a fraction whose numerator and denominator are whole numbers between 1 and 32 inclusive, thus classifying each set of ten fingerprints into one of 1024 groups. (To distinguish these groups, the fraction is *not* reduced by dividing out any common factors.) The fraction is determined by ten indicators, one for each finger, an indicator taking the value 1 when that finger has a whorl, and 0 otherwise. These indicators can be written R_t, R_i, R_m, R_r, R_l for the right hand and L_t, L_i, L_m, L_r, L_l for the left hand, where the subscripts are *t* for thumb, *i* for index finger, *m* for middle finger, *r* for ring finger and *l* for little finger. The formula for the fraction is then as follows:

$$\frac{16R_i + 8R_r + 4L_t + 2L_m + 1L_l + 1}{16R_t + 8R_m + 4R_l + 2L_i + 1L_r + 1}.$$

For example, if only the right ring finger and the left index finger have whorls, then the set of fingerprints is classified into the "9/3" group:

$$\frac{16(0)+8(1)+4(0)+2(0)+1(0)+1}{16(0)+8(0)+4(0)+2(1)+1(0)+1} = \frac{9}{3}.$$

Note that although 9/3 = 3/1, the "9/3" group is different from the "3/1" group, as the latter corresponds to having whorls only on the left middle finger.

12.1.5 Fingerprint Identification

Fingerprint identification, known as dactyloscopy, ridgeology, or hand print identification, is the process of comparing two instances of friction ridge skin impressions from human fingers or toes, or even the palm of the hand or sole of the foot, to determine whether these impressions could have come from the same individual.

The flexibility and the randomized formation of the friction ridges on skin means that no two finger or palm prints are ever exactly alike in every detail; even two impressions recorded immediately after each other from the same hand may be slightly different. Fingerprint identification, also referred to as individualization, involves an expert, or an expert computer system operating under threshold scoring rules, determining whether two friction ridge impressions are likely to have originated from the same finger or palm (or toe or sole).

In 2024, research using deep learning neural networks found contrary to "prevailing assumptions" that fingerprints from different fingers of the same person could be identified as belonging to that individual with 99.99% confidence.

Further, features used in traditional methods were nonpredictive in such identification while ridge orientation, particularly near the center of the fingerprint center provided most information.

An intentional recording of friction ridges is usually made with black printer's ink rolled across a contrasting white background, typically a white card. Friction ridges can also be recorded digitally, usually on a glass plate, using a technique called live scan. A "latent print" is the chance recording of friction ridges deposited on the surface of an object or a wall. Latent prints are invisible to the naked eye, whereas "patent prints" or "plastic prints" are viewable with the unaided eye.

Latent prints are often fragmentary and require the use of chemical methods, powder, or alternative light sources in order to be made clear. Sometimes an ordinary bright flashlight will make a latent print visible.

When friction ridges come into contact with a surface that will take a print, material that is on the friction ridges such as perspiration, oil, grease, ink, or blood, will be transferred to the surface. Factors which affect the quality of friction ridge impressions are numerous.

Pliability of the skin, deposition pressure, slippage, the material from which the surface is made, the roughness of the surface, and the substance deposited are just some of the various factors which can cause a latent print to appear differently from any known recording of the same friction ridges. Indeed, the conditions surrounding every instance of friction ridge deposition are unique and never duplicated. For these reasons, fingerprint examiners are required to undergo extensive training. The scientific study of fingerprints is called dermatoglyphics.

12.1.6 Fingerprinting Techniques

Exemplar

Exemplar prints, or known prints, is the name given to fingerprints deliberately collected from a subject, whether for purposes of enrollment in a system or when under arrest for a suspected criminal offense. During criminal arrests, a set of exemplar prints will normally include one print taken from each finger that has been rolled from one edge of the nail to the other, plain (or slap) impressions of each of the four fingers of each hand, and plain impressions of each thumb. Exemplar prints can be collected using live scan or by using ink on paper cards.

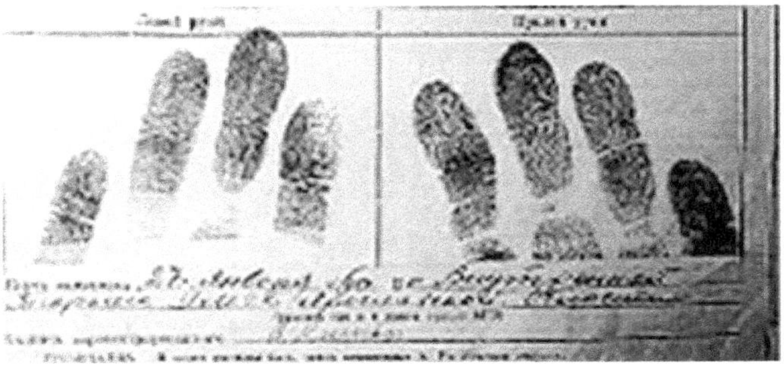

Fig 5: Exemplar Prints on Paper Using Ink

Latent

Fig 6: Barely Visible Latent Prints on a Knife

In forensic science, a partial fingerprint lifted from a surface is called a *latent fingerprint*. Moisture and grease on fingers result in latent fingerprints on surfaces such as glass. But because they are not clearly visible, their detection may require chemical development through powder dusting, the spraying of ninhydrin, iodine fuming, or soaking in silver nitrate. Depending on the surface or the material on which a latent fingerprint has been found, different methods of chemical development must be used. Forensic scientists use different techniques for porous surfaces, such as paper, and nonporous surfaces, such as glass, metal or plastic. Nonporous surfaces require the dusting process, where fine powder and a brush are used, followed by the application of transparent tape to lift the latent fingerprint off the surface.

While the police often describe all partial fingerprints found at a crime scene as latent prints, forensic scientists call partial fingerprints that are readily visible *patent prints*. Chocolate, toner, paint or ink on fingers will result in patent fingerprints. Latent fingerprints impressions that are found on soft material, such as soap, cement or plaster, are called *plastic prints* by forensic scientists.

12.1.7 Capture and Detection

Live Scan Devices

Fig 7: Fingerprint being Scanned

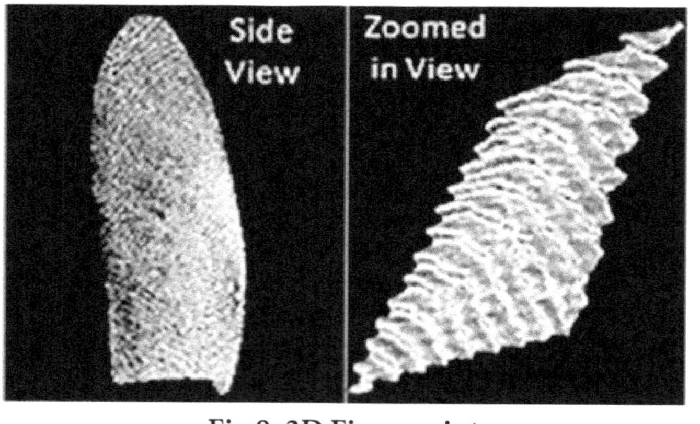

Fig 8: 3D Fingerprint

Fingerprint image acquisition is considered to be the most critical step in an automated fingerprint authentication system, as it determines the final fingerprint image quality, which has a drastic effect on the overall system performance. There are different types of fingerprint readers on the market, but the basic idea behind each is to measure the physical difference between ridges and valleys.

All the proposed methods can be grouped into two major families: solid-state fingerprint readers and optical fingerprint readers. The procedure for capturing a fingerprint using a sensor consists of rolling or touching with the finger onto a sensing area, which according to the physical principle in use (optical, ultrasonic, capacitive, or thermal) captures the difference between valleys and ridges. When a finger touches or rolls onto a surface, the elastic skin deforms. The quantity and direction of the pressure applied by the user, the skin conditions and the projection of an irregular 3D object (the finger) onto a 2D flat plane introduce distortions, noise, and inconsistencies in the captured fingerprint image. These problems result in inconsistent and non-uniform irregularities in the image. During each acquisition, therefore, the results of the imaging are different and uncontrollable. The representation of the same fingerprint changes every time the finger is placed on the sensor plate, increasing the complexity of any attempt to match fingerprints, impairing the system performance and consequently, limiting the widespread use of this biometric technology.

In order to overcome these problems, as of 2010, non-contact or touchless 3D fingerprint scanners have been developed. Acquiring detailed 3D information, 3D fingerprint scanners take a digital approach to the analog process of pressing or rolling the finger. By modelling the distance between neighboring points, the fingerprint can be imaged at a resolution high enough to record all the necessary detail.

Fingerprinting on Cadavers

The human skin itself, which is a regenerating organ until death, and environmental factors such as lotions and cosmetics, pose challenges when fingerprinting a human. Following the death of a human, the skin dries and cools. Fingerprints of dead humans may be obtained during an autopsy.

The collection of fingerprints off of a cadaver can be done in varying ways and depends on the condition of the skin. In the case of cadaver in the later stages of decomposition with dried skin, analysts will boil the skin to recondition/rehydrate it, allowing for moisture to flow back into the skin and resulting in detail friction ridges. Another method that has been used in brushing a powder, such as baby powder over the tips of the fingers. The powder will ebbed itself into the farrows of the friction ridges allowing for the lifted ridges to be seen.

Latent Fingerprint Detection

In the 1930s, criminal investigators in the United States first discovered the existence of latent fingerprints on the surfaces of fabrics, most notably on the insides of gloves discarded by perpetrators.

Since the late nineteenth century, fingerprint identification methods have been used by police agencies around the world to identify suspected criminals as well as the victims of crime. The basis of the traditional fingerprinting technique is simple. The skin on the palmar surface of the hands and feet forms ridges, so-called papillary ridges, in patterns that are unique to each individual and which do not change over time. Even identical twins (who share their DNA) do not have identical fingerprints. The best way to render latent fingerprints visible, so that they can be photographed, can be complex and may depend, for example, on the type of surfaces on which they have been left. It is generally necessary to use a "developer", usually a powder or chemical reagent, to produce a high degree of visual contrast between the ridge patterns and the surface on which a fingerprint has been deposited.

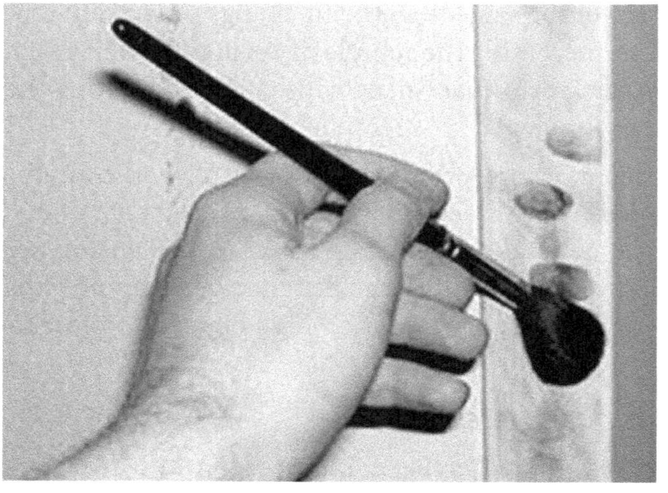

Fig 9: Use of Fine Powder and Brush to Reveal Latent Fingerprints

Fig 10: Fingerprint Dusting of a Burglary Scene

Developing agents depend on the presence of organic materials or inorganic salts for their effectiveness, although the water deposited may also take a key role. Fingerprints are typically formed from the aqueous-based secretions of the eccrine glands of the fingers and palms with additional material from sebaceous glands primarily from the forehead. This latter contamination results from the common human behaviors of touching the face and hair. The resulting latent fingerprints consist usually of a substantial proportion of water with small traces of amino acids and chlorides mixed with a fatty, sebaceous component which contains a number of fatty acids and triglycerides. Detection of a small proportion of reactive organic substances such as urea and amino acids is far from easy.

Fingerprints at a crime scene may be detected by simple powders, or by chemicals applied *in situ*. More complex techniques, usually involving chemicals, can be applied in specialist laboratories to appropriate articles removed from a crime scene. With advances in these more sophisticated techniques, some of the more advanced crime scene investigation services from around the world were, as of 2010, reporting that 50% or more of the fingerprints recovered from a crime scene had been identified as a result of laboratory-based techniques.

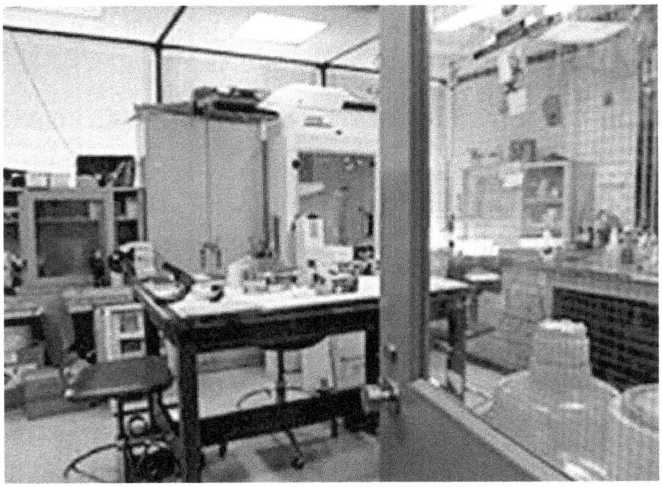

Fig 11: A City Fingerprint Identification Room

12.1.8 Forensic Laboratories

Although there are hundreds of reported techniques for fingerprint detection, many of these are only of academic interest and there are only around 20 really effective methods which are currently in use in the more advanced fingerprint laboratories around the world.

Some of these techniques, such as ninhydrin, diazafluorenone and vacuum metal deposition, show great sensitivity and are used operationally.

Some fingerprint reagents are specific, for example ninhydrin or diazafluorenone reacting with amino acids. Others such as ethyl cyanoacrylate polymerisation, work apparently by water-based catalysis and polymer growth. Vacuum metal deposition using gold and zinc has been shown to be non-specific, but can detect fat layers as thin as one molecule.

More mundane methods, such as the application of fine powders, work by adhesion to sebaceous deposits and possibly aqueous deposits in the case of fresh fingerprints. The aqueous component of a fingerprint, while initially sometimes making up over 90% of the weight of the fingerprint, can evaporate quite quickly and may have mostly gone after 24 hours. Following work on the use of argon ion lasers for fingerprint detection, a wide range of fluorescence techniques have been introduced, primarily for the enhancement of chemically developed fingerprints; the inherent fluorescence of some latent fingerprints may also be detected. Fingerprints can for example be visualized in 3D and without chemicals by the use of infrared lasers.

A comprehensive manual of the operational methods of fingerprint enhancement was last published by the UK Home Office Scientific Development Branch in 2013 and is used widely around the world. A technique proposed in 2007 aims to identify an individual's ethnicity, sex, and dietary patterns.

12.1.9 Limitations and Implications in a Forensic Context

One of the main limitations of friction ridge impression evidence regarding the actual collection would be the surface environment, specifically talking about how porous the surface the impression is on. With non-porous surfaces, the residues of the impression will not be absorbed into the material of the surface, but could be smudged by another surface. With porous surfaces, the residues of the impression will be absorbed into the surface. With both resulting in either an impression of no value to examiners or the destruction of the friction ridge impressions.

In order for analysts to correctly positively identify friction ridge patterns and their features depends heavily on the clarity of the impression. Therefore, the analysis of friction ridges is limited by clarity.

In a court context, many have argued that friction ridge identification and ridgeology should be classified as opinion evidence and not as fact, therefore should be assessed as such. Many have said that friction ridge identification is only legally admissible today because during the time when it was added to the legal system, the admissibility standards were quite low. There are only a limited number of studies that have been conducted to help confirm the science behind this identification process.

12.1.10 Crime Scene Investigations

The application of the new scanning Kelvin probe (SKP) fingerprinting technique, which makes no physical contact with the fingerprint and does not require the use of developers, has the potential to allow fingerprints to be recorded while still leaving intact material that could subsequently be subjected to DNA analysis.

A forensically usable prototype was under development at Swansea University during 2010, in research that was generating significant interest from the British Home Office and a number of different police forces across the UK, as well as internationally.

The hope is that this instrument could eventually be manufactured in sufficiently large numbers to be widely used by forensic teams worldwide.

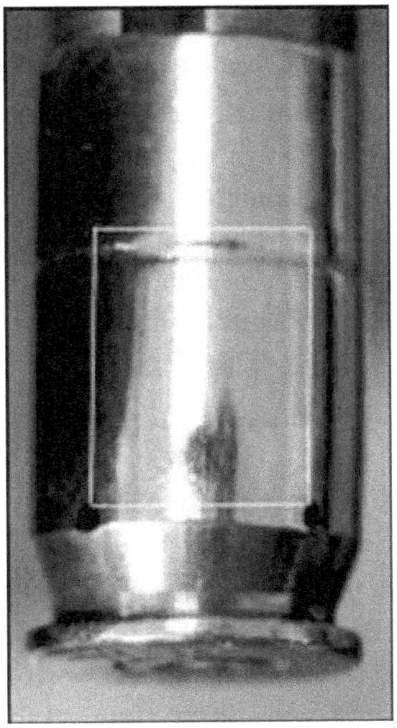

Fig 12: A Fingerprint on a Cartridge Case

12.1.11 Detection of Drug Use

The secretions, skin oils and dead cells in a human fingerprint contain residues of various chemicals and their metabolites present in the body. These can be detected and used for forensic purposes. For example, the fingerprints of tobacco smokers contain traces of cotinine, a nicotine metabolite; they also contain traces of nicotine itself. Caution should be used, as its presence may be caused by mere contact of the finger with a tobacco product. By treating the fingerprint with gold nanoparticles with attached cotinine antibodies, and then subsequently with a fluorescent agent attached to cotinine antibodies, the fingerprint of a smoker becomes fluorescent; non-smokers' fingerprints stay dark. The same approach, as of 2010, is being tested for use in identifying heavy coffee drinkers, cannabis smokers, and users of various other drugs.

12.1.12 Police Force Databases

Most American law enforcement agencies use Wavelet Scalar Quantization (WSQ), a wavelet-based system for efficient storage of compressed fingerprint images at 500 pixels per inch (ppi). WSQ was developed by the FBI, the Los Alamos National Lab, and the National Institute of Standards and Technology (NIST). For fingerprints recorded at 1000 ppi spatial resolution, law enforcement (including the FBI) uses JPEG 2000 instead of WSQ.

12.1.13 Validity

Fingerprints collected at a crime scene, or on items of evidence from a crime, have been used in forensic science to identify suspects, victims and other persons who touched a surface. Fingerprint identification emerged as an important system within police agencies in the late 19th century, when it replaced anthropometric measurements as a more reliable method for identifying persons having a prior record, often under a false name, in a criminal record repository. Fingerprinting has served all governments worldwide during the past 100 years or so to provide identification of criminals. Fingerprints are the fundamental tool in every police agency for the identification of people with a criminal history.

The validity of forensic fingerprint evidence has been challenged by academics, judges and the media. In the United States fingerprint examiners have not developed uniform standards for the identification of an individual based on matching fingerprints. In some countries where fingerprints are also used in criminal investigations, fingerprint examiners are required to match a number of *identification points* before a match is accepted. In England 16 identification points are required and in France 12, to match two fingerprints and identify an individual. Point-counting methods have been challenged by some fingerprint examiners because they focus solely on the location of particular characteristics in fingerprints that are to be matched. Fingerprint examiners may also uphold the *one dissimilarity doctrine*, which holds that if there is one dissimilarity between two fingerprints, the fingerprints are not from the same finger. Furthermore, academics have argued that the error rate in matching fingerprints has not been adequately studied and it has even been argued that fingerprint evidence has no secure statistical foundation. Research has been conducted into whether experts can objectively focus on feature information in fingerprints without being misled by extraneous information, such as context. Fingerprints can theoretically be forged and planted at crime scenes.

12.1.14 Professional Certification

Fingerprinting was the basis upon which the first forensic professional organization was formed, the International Association for Identification (IAI), in 1915. The first professional certification program for forensic scientists was established in 1977, the IAI's Certified Latent Print Examiner program, which issued certificates to those meeting stringent criteria and had the power to revoke certification where an individual's performance warranted it. Other forensic disciplines have followed suit and established their own certification programs.

12.1.15 Absence or Mutilation of Fingerprints

A very rare medical condition, adermatoglyphia, is characterized by the absence of fingerprints. Affected persons have completely smooth fingertips, palms, toes and soles, but no other medical signs or symptoms. A 2011 study indicated that adermatoglyphia is caused by the improper expression of the protein SMARCAD1. The condition has been called *immigration delay disease* by the researchers describing it, because the congenital lack of fingerprints causes delays when affected persons attempt to prove their identity while traveling. Only five families with this condition had been described as of 2011.

People with Naegeli–Franceschetti–Jadassohn syndrome and dermatopathia pigmentosa reticularis, which are both forms of ectodermal dysplasia, also have no fingerprints. Both of these rare genetic syndromes produce other signs and symptoms as well, such as thin, brittle hair.

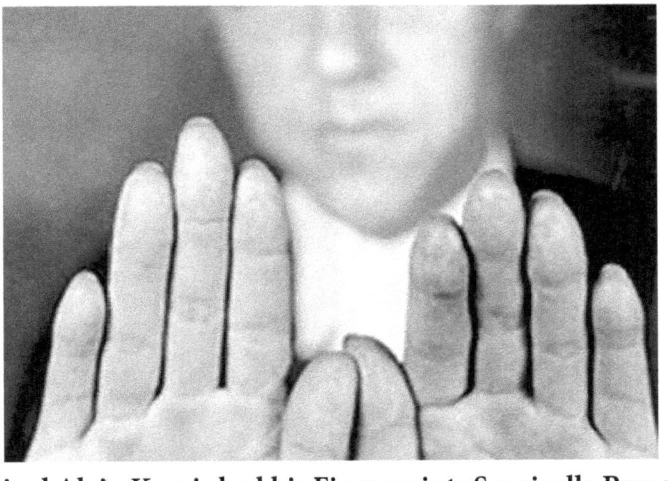

Fig 13: Criminal Alvin Karpis had his Fingerprints Surgically Removed in 1933

The anti-cancer medication capecitabine may cause the loss of fingerprints. Swelling of the fingers, such as that caused by bee stings, will in some cases cause the temporary disappearance of fingerprints, though they will return when the swelling recedes.

Since the elasticity of skin decreases with age, many senior citizens have fingerprints that are difficult to capture. The ridges get thicker; the height between the top of the ridge and the bottom of the furrow gets narrow, so there is less prominence.

Fingerprints can be erased permanently and this can potentially be used by criminals to reduce their chance of conviction. Erasure can be achieved in a variety of ways including simply burning the fingertips, using acids and advanced techniques such as plastic surgery. John Dillinger burned his fingers with acid, but prints taken during a previous arrest and upon death still exhibited almost complete relation to one another.

12.1.16 Fingerprint Verification

Fig: Ridge Ending

Fig 14: Bifurcation

Fingerprints can be captured as graphical ridge and valley patterns. Because of their uniqueness and permanence, fingerprints emerged as the most widely used biometric identifier in the 2000s.

Automated fingerprint verification systems were developed to meet the needs of law enforcement and their use became more widespread in civilian applications.

Fig 15: Short Ridge (Dot)

Despite being deployed more widely, reliable automated fingerprint verification remained a challenge and was extensively researched in the context of pattern recognition and image processing. The uniqueness of a fingerprint can be established by the overall pattern of ridges and valleys, or the logical ridge discontinuities known as minutiae. In the 2000s, minutiae features were considered the most discriminating and reliable feature of a fingerprint. Therefore, the recognition of minutiae features became the most common basis for automated fingerprint verification. The most widely used minutiae features used for automated fingerprint verification were the ridge ending and the ridge bifurcation.

Patterns

The three basic patterns of fingerprint ridges are the arch, loop, and whorl:

- **Arch:** The ridges enter from one side of the finger, rise in the center forming an arc, and then exit the other side of the finger.
- **Loop:** The ridges enter from one side of a finger, form a curve, and then exit on that same side.
- **Whorl:** Ridges form circularly around a central point on the finger.

Scientists have found that family members often share the same general fingerprint patterns, leading to the belief that these patterns are inherited.

Fingerprint Features

Features of fingerprint ridges, called *minutiae*, include:

- **Ridge Ending:** The abrupt end of a ridge
- **Bifurcation:** A single ridge dividing in two
- **Short or Independent Ridge:** A ridge that commences, travels a short distance and then ends

- **Island or Dot:** A single small ridge inside a short ridge or ridge ending that is not connected to all other ridges
- **Lake or Ridge Enclosure:** A single ridge that bifurcates and reunites shortly afterward to continue as a single ridge
- **Spur:** A bifurcation with a short ridge branching off a longer ridge
- **Bridge or Crossover:** A short ridge that runs between two parallel ridges
- **Delta:** A Y-shaped ridge meeting
- **Core:** A circle in the ridge pattern

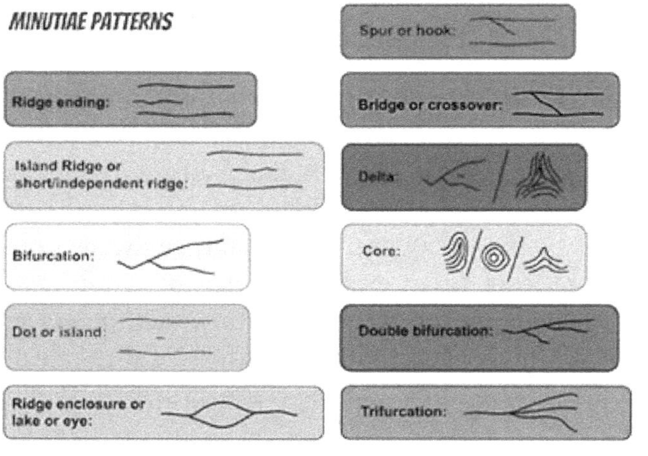

12.1.17 Fingerprint Sensors

A fingerprint sensor is an electronic device used to capture a digital image of the fingerprint pattern. The captured image is called a live scan. This live scan is digitally processed to create a biometric template (a collection of extracted features) which is stored and used for matching. Many technologies have been used including optical, capacitive, RF, thermal, piezoresistive, ultrasonic, piezoelectric, and MEMS.

- Optical scanners take a visual image of the fingerprint using a digital camera.
- Capacitive or CMOS scanners use capacitors and thus electric current to form an image of the fingerprint.
- Ultrasound fingerprint scanners use high frequency sound waves to penetrate the epidermal (outer) layer of the skin.
- Thermal scanners sense the temperature differences on the contact surface, in between fingerprint ridges and valleys.

Consumer Electronics Login Authentication

Fig 16: The Fingerprint Sensor of a Lenovo ThinkPad

Since 2000, electronic fingerprint readers have been introduced as consumer electronics security applications. Fingerprint sensors could be used for login authentication and the identification of computer users. However, some less sophisticated sensors have been discovered to be vulnerable to quite simple methods of deception, such as fake fingerprints cast in gels. In 2006, fingerprint sensors gained popularity in the laptop market. Built-in sensors in laptops, such as ThinkPad, VAIO, HP Pavilion and EliteBook laptops, and others also double as motion detectors for document scrolling, like the scroll wheel.

Two of the first smartphone manufacturers to integrate fingerprint recognition into their phones were Motorola with the Atrix 4G in 2011 and Apple with the iPhone 5S on September 10, 2013. One month after, HTC launched the One Max, which also included fingerprint recognition. In April 2014, Samsung released the Galaxy S5, which integrated a fingerprint sensor on the home button.

Following the release of the iPhone 5S model, a group of German hackers announced on September 21, 2013, that they had bypassed Apple's new Touch ID fingerprint sensor by photographing a fingerprint from a glass surface and using that captured image as verification. The spokesman for the group stated: "We hope that this finally puts to rest the illusions people have about fingerprint biometrics. It is plain stupid to use something that you can't change and that you leave everywhere every day as a security token." In September 2015, Apple included a new version of the fingerprint scanner in the iPhone home button with the iPhone 6S. The use of the Touch ID fingerprint scanner was optional and could be configured to unlock the screen or pay for mobile apps purchases. Since December 2015, cheaper smartphones with fingerprint recognition have been released, such as the $100 UMI Fair. Samsung introduced fingerprint sensors to its mid-range A series smartphones in 2014.

By 2017, Hewlett Packard, Asus, Huawei, Lenovo and Apple were using fingerprint readers in their laptops. Synaptics says the SecurePad sensor is now available for OEMs to start building into their laptops. In 2018, Synaptics revealed that their in-display fingerprint sensors would be featured on the new Vivo X21 UD smartphone. This was the first mass-produced fingerprint sensor to be integrated into the entire touchscreen display, rather than as a separate sensor.

12.1.18 Algorithms

Matching algorithms are used to compare previously stored templates of fingerprints against candidate fingerprints for authentication purposes. In order to do this either the original image must be directly compared with the candidate image or certain features must be compared.

Pre-Processing

Pre-processing enhances the quality of an image by filtering and removing extraneous noise. The minutiae-based algorithm is only effective with 8-bit gray scale fingerprint images. One reason for this is that an 8-bit gray fingerprint image is a fundamental base when converting the image to a 1-bit image with value 1 for ridges and value 0 for furrows.

This process allows for enhanced edge detection so the fingerprint is revealed in high contrast, with the ridges highlighted in black and the furrows in white. To further optimize the input image's quality, two more steps are required: minutiae extraction and false minutiae removal. The minutiae extraction is carried out by applying a ridge-thinning algorithm that removes redundant pixels of ridges. As a result, the thinned ridges of the fingerprint image are marked with a unique ID to facilitate the conduction of further operations. After the minutiae extraction, the false minutiae removal is carried out. The lack of the amount of ink and the cross link among the ridges could cause false minutiae that led to inaccuracy in fingerprint recognition process.

Pattern-Based (or Image-Based) Algorithms

Pattern based algorithms compare the basic fingerprint patterns (arch, whorl, and loop) between a previously stored template and a candidate fingerprint. This requires that the images can be aligned in the same orientation. To do this, the algorithm finds a central point in the fingerprint image and centers on that. In a pattern-based algorithm, the template contains the type, size, and orientation of patterns within the aligned fingerprint image. The candidate fingerprint image is graphically compared with the template to determine the degree to which they match.

12.2 METHODS OF LATENT PRINT DEVELOPMENT

Various techniques are employed in latent print development, each designed to enhance the visibility of prints on different types of surfaces. These methods can be broadly categorized into physical, chemical, and digital techniques.

12.2.1 Physical Methods

Physical methods rely on the application of substances that adhere to the oils and sweat present in latent prints. Common physical methods include:

- **Dusting with Powder**: One of the most widely used techniques, powder dusting involves applying a fine powder to the surface where the latent print is suspected to be. The powder adheres to the oils in the print, making it visible. Different powders, such as carbon black or fluorescent powders, can be used depending on the surface type and environmental conditions. Once developed, the print can be photographed or lifted using tape for further analysis.
- **Super Glue Fuming**: This technique, also known as cyanoacrylate fuming, is particularly effective for non-porous surfaces such as metal or plastic. The surface is placed in a fuming chamber, where super glue is heated to release cyanoacrylate vapors. These vapors adhere to the latent print, forming a visible white impression. This method is particularly useful in cases where powder dusting may not be effective due to the surface's characteristics.

12.2.2 Chemical Methods

Chemical methods involve the use of various reagents that react with the components of latent prints to produce visible results. Some common chemical techniques include:

- **Ninhydrin**: This reagent is primarily used for developing latent prints on porous surfaces, such as paper or cardboard. When ninhydrin is applied to a surface containing amino acids from sweat, it reacts to produce a purple-blue compound known as Ruhemann's purple. This method can be particularly effective in cases where prints are old or have degraded over time.

- **Silver Nitrate:** Silver nitrate is another chemical used to develop latent prints on porous surfaces. When applied, it reacts with the chloride ions in sweat to form silver chloride, which appears as a dark brown or black stain when exposed to light. This method is valuable for developing prints on surfaces that may not respond well to other chemical techniques.
- **Iodine Fuming:** Iodine is used to develop latent prints on porous surfaces. When iodine vapors come into contact with the oils in the latent print, they form a temporary brown stain. The print can be stabilized by spraying a starch solution, which reacts with the iodine, making the print more visible.

12.2.3 Digital Techniques

In recent years, advancements in technology have led to the development of digital techniques for latent print analysis. These methods often involve high-resolution imaging and software analysis to enhance and compare fingerprints.

- **Digital Imaging:** Forensic experts can use high-resolution cameras to capture detailed images of latent prints. These images can then be enhanced using digital image processing software, which allows for better visualization and analysis. Digital imaging can help in identifying subtle details that may not be visible in standard photographs.
- **Automated Fingerprint Identification Systems (AFIS):** AFIS is a computer system that stores and compares fingerprints. Once latent prints are developed, they can be scanned and entered into AFIS, which compares them against a database of known prints. This process can significantly expedite the identification of potential suspects.

12.3 CHALLENGES IN LATENT PRINT DEVELOPMENT

While latent print development is a powerful tool in forensic chemistry, it is not without challenges. Several factors can affect the successful development and analysis of latent prints.

12.3.1 Surface Type

The type of surface on which a latent print is deposited can significantly impact the effectiveness of development techniques. Non-porous surfaces, such as glass or plastic, may yield clearer prints with methods like dusting or super glue fuming.

However, porous surfaces may require chemical methods that can vary in effectiveness depending on environmental conditions, such as humidity and temperature.

12.3.2 Age and Degradation

Latent prints can degrade over time, especially if exposed to environmental factors such as heat, moisture, or chemicals. Older prints may be more challenging to develop, as the oils and sweat can evaporate or break down. In some cases, prints may be completely lost, making it impossible to establish a connection to a suspect.

12.3.3 Contamination

Contamination can pose a significant challenge in latent print analysis. The presence of dust, dirt, or other residues on a surface can obscure latent prints, making them more difficult to develop. Additionally, handling surfaces before proper evidence collection can lead to cross-contamination, complicating the analysis process.

12.4 CONCLUSION

Latent print development is an essential aspect of forensic chemistry that provides invaluable insights in criminal investigations. The ability to enhance and analyze latent prints allows forensic experts to identify individuals, link suspects to crime scenes, and corroborate evidence from other sources. Through the application of physical, chemical, and digital techniques, forensic chemists can reveal crucial evidence that supports the pursuit of justice.

Despite the challenges posed by surface types, aging, and contamination, advancements in technology and methodologies continue to improve the accuracy and effectiveness of latent print development. As forensic science evolves, the role of latent prints in investigations will remain pivotal, contributing to a deeper understanding of criminal behavior and the establishment of accountability within the justice system. The ongoing research and development in this field promise to enhance the capabilities of forensic chemists and improve the reliability of evidence used in legal proceedings.

CHAPTER 13
FIREARMS

Firearms play a significant role in forensic chemistry, particularly in criminal investigations involving gun-related crimes. The analysis of firearms and their associated evidence can provide critical insights into the circumstances surrounding a crime, helping law enforcement agencies solve cases and ensure justice is served.

13.1 UNDERSTANDING FIREARMS

Firearms are classified as weapons that use gunpowder or other propellant substances to launch projectiles. They are typically categorized into two main types: handguns and long guns. Handguns, such as revolvers and semi-automatic pistols, are designed for one-handed operation and are often used in personal defense or criminal activity. Long guns, including rifles and shotguns, are typically designed for more precise shooting and are commonly used in hunting and law enforcement scenarios.

13.1.1 Importance of Firearms in Forensic Chemistry

The significance of firearms in forensic chemistry lies in their potential to provide crucial evidence in criminal investigations. When a firearm is discharged, it produces various residues and traces that can be analyzed to determine the type of weapon used, the distance from which it was fired, and even the identity of the shooter. This information is vital for reconstructing crime scenes and establishing timelines of events.

Moreover, firearms are often linked to various forms of evidence, such as bullet casings, gunshot residue (GSR), and the projectiles themselves. The analysis of these components can reveal important details about a shooting incident, including the number of shots fired, the type of ammunition used, and the firearm's make and model. As such, firearms hold a unique position in forensic science, serving as both tools for committing crimes and as vital sources of evidence in criminal investigations.

13.2 TYPES OF EVIDENCE ASSOCIATED WITH FIREARMS

Several types of evidence are typically associated with firearms in forensic chemistry. Understanding these different forms of evidence is crucial for forensic analysts in their investigations.

13.2.1 Gunshot Residue (GSR)

Gunshot residue is a critical type of evidence that is often analyzed in firearms-related cases. When a firearm is discharged, tiny particles of gunpowder and other materials are expelled from the gun's muzzle. These particles can settle on the shooter's hands, clothing, and surrounding surfaces, providing forensic analysts with valuable information about the discharge event. GSR analysis involves the collection of samples from the suspect's hands or clothing using adhesive stubs or swabs. The collected samples are then examined under a microscope or using specialized chemical tests to identify the presence of elements such as lead, barium, and antimony, which are commonly found in GSR. The presence of these elements can indicate whether a person fired a weapon or was in close proximity to a discharged firearm.

13.2.2 Bullet Casings

Bullet casings are another essential piece of evidence in firearms investigations. When a firearm is discharged, the cartridge case is ejected, leaving behind a distinctive imprint on the casing. Forensic analysts can examine these markings, known as "striation patterns," to identify the specific firearm that discharged the bullet. Each firearm has unique characteristics that create distinctive markings on the casings it ejects. The analysis of bullet casings typically involves the use of comparison microscopes, which allow forensic experts to compare the markings on casings found at a crime scene with those from a suspect's firearm. This comparative analysis can provide compelling evidence linking a suspect to a specific crime, demonstrating that the firearm in their possession was used in the commission of a crime.

13.2.3 Bullets

Bullets themselves are also examined in forensic investigations. When a bullet strikes a target, it may deform or fragment, leaving behind identifiable marks. Forensic chemists can analyze these bullets to determine their caliber, type, and even the specific firearm that discharged them. This analysis may involve visual inspection, as well as chemical tests to determine the bullet's composition. In addition to identifying the type of ammunition used, forensic analysis of bullets can also reveal important details about the shooting incident, such as the angle of entry, the distance from which the shot was fired, and any other forensic evidence that may be present on the bullet's surface.

13.3 METHODS OF FIREARMS ANALYSIS

The analysis of firearms-related evidence involves various scientific methods that enable forensic chemists to gather and interpret data. These methods include chemical analysis, microscopy, and ballistics testing.

13.3.1 Chemical Analysis

Chemical analysis is a crucial component of firearms investigation. It involves the use of techniques such as gas chromatography and mass spectrometry to analyze gunshot residue, bullet composition, and other chemical components found in firearms-related evidence. These analyses can provide insights into the presence of specific elements, allowing forensic chemists to draw conclusions about the circumstances surrounding the discharge of a firearm.

For example, the detection of certain elements in GSR samples can provide evidence that a suspect discharged a weapon, while the composition of bullets can reveal information about the type of ammunition used in a shooting incident. Chemical analysis thus plays a critical role in establishing links between suspects, firearms, and crime scenes.

13.3.2 Microscopy

Microscopy is another essential method used in firearms analysis. Comparison microscopes enable forensic experts to examine bullet casings, bullets, and GSR samples at high magnification. This allows them to identify unique markings and striations that can be used to match evidence to a specific firearm.

Forensic analysts may also employ scanning electron microscopy (SEM) to analyze GSR particles. SEM can provide high-resolution images of the particles, allowing for a more detailed examination of their composition and morphology. This method enhances the accuracy of GSR analysis and helps confirm whether a suspect was involved in a shooting incident.

13.3.3 Ballistics Testing

Ballistics testing involves the study of the behavior and effects of projectiles, particularly in relation to firearms. Forensic ballistics experts analyze how bullets travel through various mediums and the impact they have on targets. This analysis can provide important insights into the trajectory of a bullet, helping to reconstruct shooting incidents and establish timelines.

In addition to analyzing bullet trajectories, forensic ballistics may also involve testing firearms to determine their functional characteristics. This includes examining the firearm's mechanism to ensure it operates correctly and determining its firing capabilities. Such analyses can be crucial in legal cases where the use of a specific firearm is disputed.

13.4 THE ROLE OF FORENSIC CHEMISTS IN LEGAL PROCEEDINGS

Forensic chemists specializing in firearms analysis play a critical role in the legal process. Their expertise is often called upon in criminal trials to provide testimony regarding the evidence collected and analyzed. This testimony can include explanations of the methods used in analysis, interpretations of the results, and their implications in the context of the case.

13.4.1 Expert Testimony

Forensic chemists are often required to present their findings in court as expert witnesses. This testimony is essential in helping juries and judges understand the significance of the forensic evidence. Forensic chemists must be able to communicate complex scientific concepts in a clear and understandable manner, ensuring that the legal system can effectively utilize the information obtained from firearms analysis.

13.4.2 Evidence Chain of Custody

Maintaining a proper chain of custody is crucial in forensic investigations involving firearms. Forensic chemists must ensure that all evidence is collected, stored, and analyzed following strict protocols to prevent contamination or tampering. A well-documented chain of custody not only protects the integrity of the evidence but also reinforces its admissibility in court.

13.5 CONCLUSION

Firearms play a pivotal role in forensic chemistry, providing vital evidence in criminal investigations involving gun-related crimes. Through the analysis of gunshot residue, bullet casings, and bullets themselves, forensic chemists can uncover critical information about the circumstances surrounding a shooting incident. The methods employed in firearms analysis, including chemical analysis, microscopy, and ballistics testing, enable forensic experts to draw meaningful conclusions that can significantly impact legal proceedings.

The role of forensic chemists in firearms analysis extends beyond scientific analysis; they are essential participants in the judicial process, providing expert testimony and ensuring the integrity of evidence. As forensic science continues to evolve, the analysis of firearms will remain a key area of focus, contributing to the pursuit of justice and the resolution of complex criminal cases.

CHAPTER 14
ARSONS AND EXPLOSIVES

Forensic chemistry plays a crucial role in the investigation of arsons and explosions. These incidents not only result in significant property damage and loss of life but also present unique challenges for forensic experts attempting to determine the causes and circumstances surrounding them. Understanding the chemistry behind combustible materials, as well as the methods used to analyze evidence, is essential for investigators in their quest for justice.

14.1 UNDERSTANDING ARSON

Arson is the intentional act of setting fire to property with the intent to cause damage or destruction. The motivations behind arson can vary, ranging from financial gain, revenge, vandalism, or psychological gratification. Forensic chemists are tasked with identifying accelerants, which are substances used to initiate or enhance the combustion process in a fire.

14.1.1 The Chemistry of Fire

Fire is a chemical reaction that occurs when a combustible material combines with oxygen in the presence of heat, resulting in a rapid release of energy. This process, known as combustion, can be influenced by various factors, including the type of fuel, the presence of accelerants, and environmental conditions. In forensic investigations, understanding the chemistry of combustion is essential for determining the cause of a fire and whether it was intentionally set.

Combustible Materials

Common combustible materials include wood, paper, plastics, and textiles. Each of these materials has unique chemical properties that affect how they burn. For instance, wood primarily consists of cellulose, which combusts at relatively low temperatures. In contrast, plastics, which contain hydrocarbons, can release toxic gases when burned. Forensic chemists analyze these materials to understand their combustion characteristics and identify any unusual behavior that may suggest arson.

Accelerants

Accelerants are substances that facilitate the ignition and spread of fire. Common accelerants include gasoline, kerosene, and lighter fluid. Forensic chemists are trained to identify the chemical signatures of these substances, which can provide critical evidence in arson investigations. The presence of an accelerant can indicate that a fire was intentionally set, distinguishing it from accidental fires caused by electrical malfunctions or natural occurrences.

14.2 THE ROLE OF FORENSIC CHEMISTRY IN ARSON INVESTIGATIONS

The investigation of arsons requires a systematic approach to collecting and analyzing evidence. Forensic chemists play a crucial role in this process by employing various techniques to identify accelerants and determine the fire's origin.

14.2.1 Evidence Collection

Collecting evidence at the scene of an arson is vital for successful investigations. Forensic experts must carefully document the scene, noting the location of the fire's origin, the patterns of damage, and any materials that may have been involved. Samples of suspected accelerants are collected using specialized tools, such as airtight containers, to prevent contamination or evaporation.

14.2.2 Analytical Techniques

Once evidence is collected, forensic chemists employ various analytical techniques to identify the presence of accelerants and determine the fire's cause. Some of the commonly used methods include gas chromatography, mass spectrometry, and infrared spectroscopy.

Gas Chromatography

Gas chromatography (GC) is a powerful analytical technique used to separate and identify volatile substances in a sample. In arson investigations, GC can detect the presence of accelerants in collected samples, providing crucial evidence of intentional fire-setting. The technique works by vaporizing the sample and passing it through a column, where the different components are separated based on their chemical properties. The resulting data can help forensic chemists identify specific accelerants and their concentrations.

Mass Spectrometry

Mass spectrometry (MS) is often used in conjunction with gas chromatography to provide detailed information about the molecular structure of the detected substances.

By measuring the mass-to-charge ratio of ions, forensic chemists can identify the specific compounds present in a sample, offering further evidence of the use of accelerants in an arson case.

Infrared Spectroscopy

Infrared spectroscopy (IR) is another valuable tool in forensic chemistry. This technique measures the absorption of infrared light by a sample, providing information about its molecular structure. Forensic chemists can use IR to identify specific functional groups present in accelerants, allowing them to confirm the presence of known combustible materials.

14.2.3 Scene Reconstruction

In addition to identifying accelerants, forensic chemists are also involved in reconstructing the scene of an arson. This involves analyzing burn patterns, fire spread, and the overall damage to determine the origin of the fire. Forensic experts may use mathematical models and simulations to understand the dynamics of the fire and how it interacted with the materials present at the scene.

14.3 UNDERSTANDING EXPLOSIVES

Explosives are substances that undergo rapid chemical reactions, producing gas and heat and resulting in an explosive force. Similar to arson investigations, forensic chemistry is critical in the analysis of explosives to determine their composition and origin.

14.3.1 Types of Explosives

Explosives can be classified into two main categories: primary and secondary explosives. Primary explosives are sensitive to heat, shock, and friction, making them suitable for use in detonators. Examples include lead azide and mercury fulminate. Secondary explosives, such as TNT and RDX, are more stable and require a detonator to initiate a reaction. Forensic chemists analyze these substances to identify their presence at the scene of an explosion.

14.3.2 The Chemistry of Explosives

The chemistry of explosives involves understanding the reactions that occur when these substances are detonated. Explosives release energy through rapid decomposition, which generates a shock wave and produces gases that expand rapidly. This explosion can cause significant damage to surrounding structures and pose a threat to life.

Forensic chemists study the chemical composition of explosives to identify the specific types used in an explosion. This information is essential for determining the source of the explosives and linking them to potential suspects.

14.4 FORENSIC ANALYSIS OF EXPLOSIVES

The forensic analysis of explosives involves various techniques to identify the substances involved in an explosion and reconstruct the events leading up to it.

14.4.1 Evidence Collection and Preservation

Similar to arson investigations, the collection and preservation of evidence are crucial in explosive-related cases. Forensic experts must carefully collect samples of any explosive residues, fragments, and other materials from the blast site. Proper handling and storage techniques are essential to prevent contamination and degradation of evidence.

14.4.2 Analytical Techniques for Explosives

Forensic chemists utilize various analytical methods to identify and characterize explosives. Techniques such as gas chromatography-mass spectrometry (GC-MS), ion mobility spectrometry (IMS), and Raman spectroscopy are commonly employed.

Gas Chromatography-Mass Spectrometry (GC-MS)

GC-MS is a widely used technique for the analysis of explosive residues. It combines the separation capabilities of gas chromatography with the identification capabilities of mass spectrometry, allowing forensic chemists to detect and identify specific explosive compounds. This method is particularly effective for analyzing volatile organic compounds released during explosions.

Ion Mobility Spectrometry (IMS)

Ion mobility spectrometry is a rapid detection method often used in field settings. IMS can identify explosive residues by measuring the time it takes for ions to travel through a gas under an electric field. This technique is highly sensitive and can provide real-time results, making it valuable for on-site investigations.

Raman Spectroscopy

Raman spectroscopy is another technique used for explosive analysis. It involves shining a laser on a sample and measuring the scattered light to identify the molecular composition. This method can detect a wide range of explosives, including those that are difficult to analyze using other techniques.

14.5 LEGAL IMPLICATIONS OF FORENSIC FINDINGS

The findings of forensic investigations into arsons and explosives have significant legal implications. The analysis of accelerants, explosive residues, and fire patterns can provide critical evidence in criminal cases, helping to establish the intent and actions of suspects.

14.5.1 Expert Testimony in Court

Forensic chemists often serve as expert witnesses in court, providing testimony regarding their analyses and findings. Their expertise is essential for helping juries and judges understand the scientific basis of the evidence presented. Effective communication of complex scientific concepts is crucial in ensuring that the findings are accurately interpreted within the legal context.

14.5.2 Chain of Custody

Maintaining a proper chain of custody for evidence is critical in forensic investigations. Forensic chemists must ensure that all evidence collected is properly documented, stored, and handled throughout the investigation process. A clear chain of custody reinforces the integrity of the evidence and its admissibility in court.

14.6 CONCLUSION

Arsons and explosives present significant challenges for forensic chemists, who play a vital role in investigating these complex incidents. Understanding the chemistry behind combustion and explosive reactions is essential for identifying accelerants, analyzing residues, and reconstructing events. The methods employed in forensic investigations, including gas chromatography, mass spectrometry, and infrared spectroscopy, enable forensic experts to gather crucial evidence that can impact legal proceedings.

As forensic science continues to evolve, the importance of arson and explosives analysis remains paramount. Forensic chemists are essential in the pursuit of justice, ensuring that the chemical aspects of these incidents are thoroughly examined and understood in the context of the law. Their work not only aids in solving crimes but also contributes to a broader understanding of fire dynamics and explosive behavior, ultimately enhancing public safety and security.

CHAPTER 15

CBRNE CHEMICAL, BIOLOGICAL, RADIOLOGICAL, NUCLEAR AND EXPLOSIVES

CBRNE, which stands for Chemical, Biological, Radiological, Nuclear, and Explosives, encompasses a range of hazardous materials that pose significant threats to public safety and security. In the field of forensic chemistry, the analysis of CBRNE agents is critical for understanding incidents involving these substances, whether they are accidental releases, criminal acts, or acts of terrorism. Forensic chemists play a vital role in identifying, quantifying, and characterizing these hazardous materials, providing crucial evidence that informs investigations and legal proceedings.

15.1 UNDERSTANDING CBRNE THREATS

CBRNE agents are categorized based on their distinct properties and the threats they pose. Each category presents unique challenges for forensic chemists in terms of detection, analysis, and remediation.

15.1.1 Chemical Agents

Chemical agents refer to toxic substances that can cause harm to human health, the environment, or both.

These agents can be further divided into various classes, including nerve agents, blister agents, choking agents, and incapacitating agents. Examples include sarin, mustard gas, and chlorine gas. In forensic chemistry, the identification and analysis of chemical agents are paramount, particularly in cases of chemical warfare or terrorist attacks.

Detection and Analysis of Chemical Agents

Forensic chemists employ various analytical techniques to detect and characterize chemical agents.

Techniques such as gas chromatography-mass spectrometry (GC-MS), liquid chromatography-mass spectrometry (LC-MS), and infrared spectroscopy are commonly used. These methods allow for the identification of specific chemical compounds, helping to determine their source and potential impact on public safety.

15.1.2 Biological Agents

Biological agents include pathogens and toxins that can cause diseases in humans, animals, or plants. Examples of biological agents include bacteria (e.g., anthrax), viruses (e.g., smallpox), and toxins (e.g., botulinum toxin). The intentional release of biological agents is a major concern for public health and safety, particularly in bioterrorism scenarios.

Forensic Analysis of Biological Agents

The analysis of biological agents often involves the collection and examination of samples from potential exposure sites. Forensic chemists utilize techniques such as polymerase chain reaction (PCR) and enzyme-linked immunosorbent assay (ELISA) to detect and identify specific pathogens. These methods enable rapid identification of biological threats and aid in the implementation of appropriate public health responses.

15.1.3 Radiological Agents

Radiological agents encompass radioactive materials that can pose health risks through exposure or contamination. These agents may be used in dirty bombs or other malicious acts aimed at instilling fear and causing panic. Forensic chemists play a crucial role in the detection and analysis of radiological materials in incidents involving radiation exposure or contamination.

Detection and Analysis of Radiological Materials

The detection of radiological agents often requires specialized equipment, such as gamma spectrometers and dosimeters. Forensic chemists work closely with radiation safety experts to measure radiation levels and identify the specific isotopes present in a given sample. The ability to accurately characterize radiological agents is essential for determining the extent of contamination and devising effective response strategies.

15.1.4 Nuclear Agents

Nuclear agents refer to materials that can be used in nuclear weapons or devices. These agents include fissile materials, such as uranium and plutonium, and can result in catastrophic consequences if used maliciously. The analysis of nuclear agents is critical in preventing nuclear terrorism and ensuring national security.

Forensic Analysis of Nuclear Materials

Forensic chemists utilize advanced techniques to analyze nuclear materials, including mass spectrometry and radiochemical analysis. These methods enable the identification of isotopes and their concentrations, providing insights into the origin and potential use of nuclear materials in criminal activities. Additionally, forensic experts work with governmental agencies to monitor and secure nuclear facilities and materials.

15.1.5 Explosives

Explosives encompass a wide range of materials that can cause explosions or detonations. These materials can be used in criminal activities, terrorist acts, or accidental explosions. Forensic chemists are tasked with identifying explosive residues and determining the type of explosive used in an incident.

Analysis of Explosive Residues

The analysis of explosive residues is conducted using techniques such as gas chromatography-mass spectrometry (GC-MS) and ion mobility spectrometry (IMS). These methods allow forensic chemists to detect trace amounts of explosives at crime scenes, linking suspects to explosive-related incidents. Understanding the chemical composition of explosives is critical for determining the potential motive and methods used in criminal activities.

15.2 METHODOLOGIES IN CBRNE FORENSIC ANALYSIS

Forensic analysis of CBRNE agents involves a systematic approach to evidence collection, analysis, and interpretation. Each category of agent requires specialized techniques and expertise.

15.2.1 Evidence Collection

The collection of evidence is a critical first step in CBRNE investigations. Forensic chemists must follow strict protocols to ensure the integrity of samples. Evidence collection often involves using specialized containers, ensuring proper labeling, and maintaining a clear chain of custody to preserve the validity of the evidence.

15.2.2 Analytical Techniques

Forensic chemists employ a range of analytical techniques tailored to the specific characteristics of CBRNE agents. Techniques may include:

- **Gas Chromatography-Mass Spectrometry (GC-MS):** Widely used for analyzing volatile organic compounds, including chemical agents and explosive residues.
- **Liquid Chromatography-Mass Spectrometry (LC-MS):** Effective for analyzing non-volatile compounds, including biological toxins and drugs.
- **Polymerase Chain Reaction (PCR):** A powerful technique for amplifying and detecting specific DNA sequences, enabling the identification of biological agents.

- **Enzyme-Linked Immunosorbent Assay (ELISA):** A method used for detecting proteins and toxins, particularly in biological agent analysis.
- **Gamma Spectrometry:** Employed for the detection and characterization of radiological materials by measuring gamma radiation emissions.

15.2.3 Interpretation of Results

Interpreting the results of forensic analyses is a complex process that requires expertise in chemistry, biology, and radiological sciences.

Forensic chemists must evaluate the significance of the findings in the context of the incident, considering factors such as the concentration of the agents, potential exposure routes, and the likely impact on public safety.

15.3 LEGAL IMPLICATIONS OF CBRNE FORENSIC FINDINGS

The findings of forensic analyses of CBRNE agents have significant legal implications, particularly in cases involving terrorism, criminal activity, or public health threats.

15.3.1 Expert Testimony in Court

Forensic chemists often serve as expert witnesses in court, providing testimony regarding the methods and results of their analyses.

Their expertise is essential for helping juries and judges understand the scientific basis of the evidence presented. Clear and effective communication of complex scientific concepts is crucial for ensuring that the findings are accurately interpreted within the legal context.

15.3.2 Chain of Custody and Evidence Integrity

Maintaining a proper chain of custody for evidence is critical in CBRNE investigations. Forensic chemists must ensure that all evidence collected is properly documented, stored, and handled throughout the investigation process. A clear chain of custody reinforces the integrity of the evidence and its admissibility in court, ensuring that forensic findings can be relied upon during legal proceedings.

15.4 CHALLENGES AND FUTURE DIRECTIONS

The analysis of CBRNE agents presents numerous challenges for forensic chemists, including the need for specialized training, access to advanced analytical technologies, and the complexity of interpreting results. Additionally, the evolving landscape of terrorism and criminal activity requires continuous adaptation and innovation in forensic methodologies.

15.4.1 Emerging Technologies

Advancements in analytical techniques, such as portable mass spectrometry and advanced imaging methods, hold promise for enhancing the speed and accuracy of CBRNE analyses. These technologies enable rapid on-site testing and analysis, improving response times in critical situations.

15.4.2 Interdisciplinary Collaboration

The complexities of CBRNE analysis necessitate collaboration among forensic chemists, law enforcement, public health officials, and other stakeholders. Interdisciplinary cooperation fosters a comprehensive approach to addressing CBRNE threats, enhancing the effectiveness of investigations and public safety initiatives.

15.5 CONCLUSION

CBRNE agents represent a significant challenge for forensic chemistry, requiring specialized knowledge and techniques for their detection, analysis, and interpretation. Forensic chemists play a vital role in investigating incidents involving chemical, biological, radiological, nuclear, and explosive materials, providing critical evidence that informs legal proceedings and public safety responses.

As the landscape of threats continues to evolve, the need for advanced forensic methodologies, interdisciplinary collaboration, and ongoing education in CBRNE analysis will be paramount. By staying at the forefront of technological advancements and fostering cooperation among diverse stakeholders, forensic chemists can effectively contribute to the mitigation of CBRNE threats, ultimately enhancing public safety and security.

CHAPTER 16
ENVIRONMENTAL FORENSICS

Environmental forensics is a multidisciplinary field that integrates principles from forensic science, environmental science, and law to investigate environmental contamination and its impacts. This discipline aims to identify the sources of pollution, assess the extent of contamination, and understand the pathways through which pollutants enter the environment. The increasing concern about environmental degradation and public health issues has elevated the importance of environmental forensics in regulatory frameworks, litigation, and remediation efforts.

16.1 UNDERSTANDING ENVIRONMENTAL FORENSICS

Environmental forensics focuses on the investigation of contaminated sites and the analysis of pollutants to establish the origins and extent of environmental damage. It involves a systematic approach to gather evidence and provide scientific data that can support legal cases related to environmental crimes, liability disputes, and regulatory compliance. The ultimate goal is to ensure accountability for pollution and facilitate the restoration of contaminated environments.

16.1.1 Key Objectives of Environmental Forensics

The primary objectives of environmental forensics include:

- **Identifying Contaminants**: Forensic scientists work to identify the specific pollutants present at a site, such as heavy metals, hydrocarbons, pesticides, or radioactive materials. This identification process is crucial for determining the potential risks to human health and the environment.
- **Source Identification**: A critical aspect of environmental forensics is determining the source of contamination. This may involve analyzing chemical signatures, historical records, and environmental conditions to trace pollutants back to their origin, whether from industrial activities, waste disposal, or natural sources.
- **Extent of Contamination**: Assessing the spatial and temporal extent of contamination is vital for understanding the severity of the issue. Environmental forensics employs various sampling and analytical

techniques to characterize the distribution of pollutants across different media, such as soil, water, and air.

- **Legal Support**: Environmental forensic investigations often support legal actions, including litigation related to environmental violations, liability claims, and regulatory compliance. The findings of forensic analyses can serve as critical evidence in court and help establish accountability for environmental harm.

16.2 METHODOLOGIES IN ENVIRONMENTAL FORENSICS

Environmental forensics employs a wide range of methodologies to investigate contamination. These methodologies are often interdisciplinary, drawing from chemistry, biology, geology, and environmental science.

16.2.1 Sampling Techniques

Effective sampling is essential for obtaining representative data about contaminants. Environmental forensics uses several sampling strategies, including:

- **Soil Sampling**: Soil samples are collected to assess the presence of contaminants, such as heavy metals, hydrocarbons, and pesticides. Sampling depth and location are critical to ensure that the samples reflect the extent of contamination.
- **Water Sampling**: Surface water and groundwater samples are analyzed for pollutants. This involves collecting water samples at various depths and locations to understand the distribution of contaminants in aquatic environments.
- **Air Sampling**: Air quality assessments are conducted to measure the concentration of airborne pollutants. Techniques such as passive sampling and active sampling are used to capture volatile organic compounds (VOCs) and particulate matter.

16.2.2 Analytical Techniques

The analysis of environmental samples is crucial for identifying contaminants and understanding their behavior in the environment. Common analytical techniques used in environmental forensics include:

- **Gas Chromatography-Mass Spectrometry (GC-MS)**: This technique is widely used for analyzing organic compounds in environmental samples. GC-MS can identify and quantify pollutants such as VOCs, pesticides, and industrial solvents.
- **Inductively Coupled Plasma Mass Spectrometry (ICP-MS)**: ICP-MS is employed to analyze heavy metals and trace elements in soil and

water samples. This technique provides precise measurements of metal concentrations and helps assess potential health risks.

- **Infrared Spectroscopy**: Infrared spectroscopy is used to identify functional groups and chemical bonds in organic compounds. It is particularly useful for characterizing complex mixtures of pollutants.

16.2.3 Data Interpretation

Interpreting the data obtained from environmental analyses is a crucial step in environmental forensics. Forensic scientists evaluate the analytical results to draw conclusions about the nature and source of contamination. This may involve:

- **Chemical Fingerprinting**: Identifying unique chemical signatures or patterns associated with specific sources of pollution. This technique helps link contaminants to their origins, such as industrial processes or agricultural activities.
- **Statistical Analysis**: Employing statistical methods to assess the significance of contamination levels and identify correlations between pollutants and potential sources. This analysis aids in understanding the extent of environmental impact.

16.3 APPLICATIONS OF ENVIRONMENTAL FORENSICS

Environmental forensics has numerous applications across various sectors, including regulatory compliance, litigation, and environmental remediation.

16.3.1 Regulatory Compliance

Government agencies and environmental regulators often rely on environmental forensics to ensure compliance with environmental laws and regulations. Forensic investigations can identify violations, assess the severity of contamination, and inform remediation efforts. Environmental forensics also aids in developing regulatory frameworks and standards for pollution prevention and control.

16.3.2 Litigation Support

In legal cases involving environmental contamination, forensic scientists provide expert testimony and scientific evidence. Their analyses can establish causation and liability, helping to resolve disputes between parties involved in pollution cases. Environmental forensics plays a pivotal role in environmental lawsuits, enabling affected communities to seek justice and compensation for damages.

16.3.3 Remediation Planning

Environmental forensics informs remediation strategies by providing crucial data about the extent and nature of contamination. Forensic investigations help identify the most effective methods for cleaning up contaminated sites, such as soil excavation, bioremediation, or chemical treatments. Understanding the sources and pathways of pollutants enables practitioners to develop targeted and efficient remediation plans.

16.4 CHALLENGES IN ENVIRONMENTAL FORENSICS

Despite its significance, environmental forensics faces several challenges that can complicate investigations and analyses.

16.4.1 Complex Mixtures of Contaminants

Contaminated sites often contain complex mixtures of pollutants, making identification and quantification challenging. The presence of multiple contaminants can interfere with analytical techniques and complicate data interpretation. Forensic scientists must employ advanced methodologies and extensive experience to navigate these complexities effectively.

16.4.2 Temporal and Spatial Variability

Environmental conditions can vary significantly over time and space, influencing the behavior and distribution of pollutants. Changes in weather, land use, and biological activity can affect contaminant levels, complicating the assessment of environmental impacts. Forensic investigations must account for these variations to provide accurate and reliable data.

16.4.3 Legal and Ethical Considerations

Environmental forensics operates within a legal framework that may impose constraints on evidence collection and analysis. Ensuring compliance with legal requirements while conducting thorough investigations can be challenging. Additionally, ethical considerations related to environmental justice and community engagement must be addressed, particularly in cases involving vulnerable populations.

16.5 THE FUTURE OF ENVIRONMENTAL FORENSICS

As environmental issues continue to evolve, so too will the field of environmental forensics. Advances in analytical technologies, data analysis techniques, and interdisciplinary collaboration will shape the future of this discipline.

16.5.1 Technological Innovations

Emerging technologies, such as remote sensing, nanotechnology, and machine learning, hold promise for enhancing environmental forensic investigations. These innovations can improve data collection, analysis, and interpretation, enabling more effective assessments of contamination and its sources.

16.5.2 Interdisciplinary Collaboration

The complexity of environmental issues necessitates collaboration among experts from various fields, including chemistry, biology, geology, law, and public policy. Interdisciplinary partnerships will enhance the effectiveness of environmental forensics, fostering comprehensive solutions to environmental challenges.

16.6 CONCLUSION

Environmental forensics is a vital field that addresses the pressing issues of environmental contamination and its implications for public health and safety. Through systematic investigations, advanced analytical techniques, and expert legal support, environmental forensic scientists play a crucial role in identifying pollution sources, assessing contamination levels, and informing remediation efforts.

As the challenges posed by environmental degradation continue to grow, the importance of environmental forensics in regulatory compliance, litigation, and public safety will only increase. By leveraging technological innovations and fostering interdisciplinary collaboration, the field of environmental forensics can continue to evolve, contributing to the protection of our environment and the well-being of communities worldwide.

BIBLIOGRAPHY

Applied Forensic Chemistry *Authors*: A. K. Gupta, J. K. Das *Publisher*: Springer *Year*: 2018

Forensic Analysis on the Cutting Edge: New Methods for Trace Evidence *Editors*: David C. W. Muirhead, James M. Muirhead *Publisher*: Wiley *Year*: 2016

Forensic Analytical Chemistry *Authors*: Mark M. K. Chien, Tom V. G. P. W. De Ruyter *Publisher*: Wiley *Year*: 2015

Forensic Chemistry and Analysis *Editors*: L. J. Martin, K. H. P. Lee *Publisher*: Elsevier *Year*: 2014

Forensic Chemistry and Crime Scene Investigation *Authors*: M. B. A. Davis, K. J. E. Theobald *Publisher*: CRC Press *Year*: 2019

Forensic Chemistry and its Applications *Authors*: R. H. Taylor, T. L. Wright *Publisher*: Wiley *Year*: 2021

Forensic Chemistry *Authors*: Suzanne Bell *Publisher*: Prentice Hall *Year*: 2012

Forensic Chemistry in the 21st Century *Editors*: L. R. Emmert, D. A. Stalnaker *Publisher*: Wiley *Year*: 2018

Forensic Chemistry: A Comprehensive Overview *Authors*: C. E. Bigelow, A. S. Theis *Publisher*: Elsevier *Year*: 2021

Forensic Chemistry: A Laboratory Manual *Authors*: Steven B. Smith, Ronald K. Siegel *Publisher*: Wiley *Year*: 2014

Forensic Chemistry: A Modern Introduction *Authors*: Steven B. Smith *Publisher*: Springer *Year*: 2013

Forensic Chemistry: A Modern Perspective *Authors*: R. C. M. Desai, S. K. Y. Acharya *Publisher*: Academic Press *Year*: 2021

Forensic Chemistry: A Practical Guide *Authors*: Mark J. Zavaleta *Publisher*: Academic Press *Year*: 2018

Forensic Chemistry: A Practical Guide *Authors*: Steven B. Smith, Ronald K. Siegel *Publisher*: Academic Press *Year*: 2017

Forensic Chemistry: An Introduction *Authors*: Bruce A. M. Silverman, John J. Ryckman *Publisher*: Academic Press *Year*: 2021

Forensic Chemistry: An Overview *Authors*: James E. Girard, David S. Coons *Publisher*: Elsevier *Year*: 2017

Forensic Chemistry: Analysis and Application *Authors*: J. K. Smith, M. L. Hartman *Publisher*: Springer *Year*: 2018

Forensic Chemistry: Emerging Trends *Authors*: K. B. Johnson, J. P. Simons *Publisher*: Elsevier *Year*: 2017

Forensic Chemistry: Fundamentals and Applications *Authors*: Jay A. Siegel, Andrew K. Kapp, and Steven L. Goodrow *Publisher*: Wiley *Year*: 2010

Forensic Chemistry: Fundamentals and Applications *Authors*: William J. Tilstone, Howard E. Adger *Publisher*: Routledge *Year*: 2016

Forensic Chemistry: Investigative Techniques *Authors*: A. L. Thacker, B. C. M. Miller *Publisher*: Wiley *Year*: 2020

Forensic Chemistry: Modern Approaches *Authors*: S. P. H. Omnia, T. B. Sharma *Publisher*: CRC Press *Year*: 2021

Forensic Chemistry: Principles and Practice *Authors*: William L. Myers *Publisher*: Pearson *Year*: 2015

Forensic Chemistry: The Role of Chemistry in Crime Scene Investigation *Authors*: N. R. Flanagan, B. T. K. O'Neill *Publisher*: Elsevier *Year*: 2020

Forensic Chemistry: Theory and Practice *Authors*: Michel A. R. Bertolotti, Jerry D. Smith *Publisher*: Elsevier *Year*: 2018

Forensic Drug Chemistry *Authors*: K. L. Ring, J. R. Kreitman *Publisher*: CRC Press *Year*: 2015

Forensic Science: A Comprehensive Approach *Authors*: L. C. A. P. Schmidt, E. M. McDonnell *Publisher*: Elsevier *Year*: 2020

Forensic Science: A Comprehensive Introduction *Authors*: R. S. R. Alford, M. G. Whiting *Publisher*: Wiley *Year*: 2020

Forensic Science: An Introduction to Criminalistics *Authors*: Stuart H. James, Jon J. Nordby *Publisher*: Pearson *Year*: 2009

Forensic Science: An Introduction to the Forensic Sciences *Authors*: William G. McCarthy, Robert G. D'Antonio *Publisher*: Prentice Hall *Year*: 2014

Forensic Science: From Crime Scene to Court *Authors*: I. A. Garfinkel, L. F. Powell *Publisher*: Prentice Hall *Year*: 2017

Forensic Science: From the Crime Scene to the Crime Lab *Authors*: Richard Saferstein *Publisher*: Prentice Hall *Year*: 2017

Forensic Science: Fundamentals and Investigations *Authors*: Anthony J. Bertino, Patricia B. Bertino *Publisher*: Pearson *Year*: 2018

Forensic Science: Principles and Applications *Authors*: M. D. D. T. Morgan, P. G. B. Cross *Publisher*: Springer *Year*: 2015

Forensic Science: Principles and Applications *Authors*: M. J. D. McKinney, R. K. Thompson *Publisher*: Springer *Year*: 2015

Forensic Science: The Basics *Authors*: R. D. M. Kumar, S. N. J. Luong *Publisher*: Wiley *Year*: 2016

Forensic Science: Theory and Practice *Authors*: Paul M. C. P. Dufour, Gerald L. G. Sharman *Publisher*: CRC Press *Year*: 2015

Forensic Toxicology: Principles and Concepts *Authors*: Steven B. Karch *Publisher*: CRC Press *Year*: 2014

Forensic Toxicology: Principles and Techniques *Authors*: P. C. D. J. White, M. N. McLean *Publisher*: CRC Press *Year*: 2016

Handbook of Forensic Analytical Chemistry *Authors*: T. J. Clow, D. S. Kline *Publisher*: Wiley *Year*: 2018

Handbook of Forensic Drug Analysis *Authors*: Don M. Johnson, Steven K. T. Hurst *Publisher*: CRC Press *Year*: 2015

Introduction to Forensic Chemistry *Authors*: Lawrence Kobilinsky *Publisher*: Wiley *Year*: 2019

Introduction to Forensic Science *Authors*: Edward J. Imwinkelried, David H. Kaye *Publisher*: LexisNexis *Year*: 2014

Modern Forensic Science *Authors*: K. E. S. V. R. Jagannathan, C. K. N. Sang *Publisher*: Wiley *Year*: 2021

Modern Forensic Toxicology *Authors*: Thomas E. J. M. O'Malley, Judith A. Lesher *Publisher*: CRC Press *Year*: 2015

Practical Forensic Chemistry *Authors*: Ian D. Ward, R. L. Borkowski *Publisher*: Springer *Year*: 2017

Practical Guide to Forensic Chemistry *Authors*: H. C. L. Ramos, J. M. P. G. Jones *Publisher*: Academic Press *Year*: 2016

Principles and Practice of Forensic Chemistry *Editors*: John M. S. Finlay, Angela J. Guarino *Publisher*: Wiley *Year*: 2019

Principles of Forensic Science *Authors*: R. S. Mehta *Publisher*: Vikas Publishing House *Year*: 2012

INDEX

W

X

Y

Z

www.ingramcontent.com/pod-product-compliance
Lightning Source LLC
LaVergne TN
LVHW062120200926
848484LV00007B/1724